New Unesco source book for science teaching

New Unesco source book for science teaching

Unesco Paris 1973

Published by the United Nations
Educational, Scientific and Cultural Organization
7 place de Fontenoy, 75700 Paris
Printed by Firmin-Didot S.A.
Mesnil-sur-l'Estrée (Eure)

ISBN 92-3-101058-1

French édition: ISBN 92-3-201058-1

Preface

The *New Unesco Source Book for Science Teaching* has been prepared with the intention of bringing the *Unesco Source Book for Science Teaching* up to date, and of providing a broader coverage of scientific material likely to be included in introductory science courses. Provision for the new edition was made by the General Conference of Unesco at its fifteenth session in 1968, following requests by Member States. The revision was co-ordinated at the Science Teaching Center of the University of Maryland, U.S.A., under the general editorship of Dr. J. David Lockard, Director of the Science Teaching Center and its International Clearing House on Science and Mathematics Curricular Developments. Additional members of the revision team were Drs. Alfred de Vito, J. Dudley Herron, Ralph W. Lefler, Robert W. Menefee and Wayne Taylor. Final revision of the manuscript was carried out by Dr. H. Ibstedt, J. Kent and E. G. Smith.

In preparation for the revision, large-scale comment and feed-back from users of the previous editions of the *Unesco Source Book for Science Teaching* was collected by the World Confederation of Organizations of the Teaching Profession (WCOTP). Teachers' organizations and professional associations were invited to contribute suggestions for improvement, and a special study was conducted by the Zambia Association for Science Education. Subsequently, a meeting was convened under the auspices of the WCOTP at which guidelines for the revision were drawn up.

The history of the *Unesco Source Book for Science Teaching* goes back to the close of the Second World War. At that time, Unesco sponsored a small volume entitled *Suggestions for Science Teachers in Devastated Countries,* written by J. P. Stephenson (ex-science master at the City of London School, member of the Royal Society Committee for Cooperation with Unesco, United Kingdom). This book, while proving useful for devastated areas, was a phenomenal success in regions where previously there had been little or no equipment for practical science teaching. In 1956 the book was considerably expanded, especially by the incorporation of suggestions from Unesco science teaching field experts for making simple equipment and for carrying out experiments using locally available materials. It became the first edition of the *Unesco Source Book for Science Teaching.*

A second edition was produced in 1962, and since then the book has been reprinted 24 times and has been translated into 30 languages. To date, almost 750,000 copies have been sold.

To give credit to all who have contributed to the making of this volume would be quite impossible. Much of the material included has its origin buried deep in the past and has come to be a part of the common heritage of science teachers everywhere. In addition to the work of J. P. Stephenson and of Dr. J. David Lockard and his collaborators, previously referred to, special acknowledgement is made of the work of the many individuals and groups who have contributed in various ways to this edition. Recognition is also due to those whose names are recorded in the prefaces to previous editions.

Acknowledgements

Many helpful ideas which were modified and adapted for use in the *Source Book* are contained in works of the following writers and publishers: R. and M. Buchsbaum, A. D. Bulman, Louis T. Cox Jr., Alfred E. Friedl, Paul D. Merrick, Alberta Whitfield, R. Kudo, R. Sund, L. Trowbridge, Henry Holt· and Co., Charles E. Merrill Publishing Company, National Science Teachers Association, U.S.A., Association for Science Education, U.K., and University of Chicago Press.

Other science method source books have, of course, been referred to and grateful acknowledgement is made to : the *Source book for Elementary Science* by Hone, Joseph and Victor; *A Sourcebook for the Physical Sciences* by Joseph, Brandwein, Morholt, Pollack and Castka; *A Sourcebook for the Biological Sciences* by Morholt, Brandwein, and Joseph—all published by Harcourt Brace Jovanovich Inc.; the *Geology and Earth Sciences Sourcebook*, published by the American Geological Institute.

Ideas suggested by various curriculum projects such as the Nuffield Projects, U.K., and the U.S.A. National Science Foundation sponsored ones (i.e. ISCS, BSCS, PSSC) and those of the State Departments of Education in the U.S.A. have been most .fruitful, and thanks are due to all these sources.

The detailed star chart for inter-tropic areas was produced by Mr. H. A. Diamand, Unesco expert, in People's Republic of the Congo, especially for this publication. All the illustrations have been drawn by Miss Dominique Bazin and Mr. Paolo Moriggia.

Contents

Introduction

The search for a fuller understanding of the phenomena about us, which is the process of scientific investigation, is pursued by scientists in every part of the world. Just as science itself is universal, so the quest for improved methods of teaching science is also universal. The *New Unesco Source Book for Science Teaching* consists of ideas contributed by teachers all over the world for the use of common and widely available resources and materials in science teaching. The book is addressed to teachers, particularly teachers of science in elementary schools and the lower classes of secondary schools, and to teachers-in-training.

If science is to be learned effectively, it must be experienced. Science is so close to the life of every girl and boy that no teacher need ever be without first-hand materials for the study of science. The world within us, beneath us, around us and above us, in any part of the globe, provides an inexhaustible supply of phenomena which can be used as the subject-matter of science teaching, and of materials which can be used to construct scientific equipment and teaching aids.

The *New Unesco Source Book for Science Teaching* is intended to serve as a source of ideas for the devising of simple scientific activities, investigations and experiments which can be carried out by pupils themselves, and for the construction of simple science equipment, using materials available in the particular locality where the science teaching is taking place. As local resources differ widely within a particular country, as well as from one country to another, it is anticipated that each teacher will draw from it material appropriate to the needs of his or her own pupils and particular teaching situation.

The *New Unesco Source Book for Science Teaching* may also be used by groups of pupils engaged, for example, in science club activities, or by individual pupils wishing to carry out personal science activities and investigations. However, it is assumed that such activities will be carried out under the general guidance and supervision of a teacher to enable the pupil to derive the utmost benefit from his experiences and findings, and also, in many cases, because of necessary safety precautions. For this reason, the book is not addressed directly to pupils.

The new, revised edition has been prepared with the intention of bringing the *Unesco Source Book for Science Teaching* up to date and of reflecting modern approaches to science teaching at the elementary and lower-secondary levels. In view of the far-reaching nature of recent advances in the approach and methodology of science teaching, no attempt has been made to include within this single volume suggestions concerning comprehensive teaching strategies. These will be contained in a companion volume, the *Unesco Handbook for Science Teachers*, which will also cover aspects of the learning process in children, and sociological considerations as they relate to the activities of practising school science teachers.

If pupils are to have the breadth of view needed to tackle the problems arising from the application of science to their everyday lives, they should be equipped to do so through science teaching which is broadly based. Such teaching draws carefully selected material from the whole range of the sciences, including earth and space sciences and interdisciplinary areas. To this end, the scope of the *New Unesco Source Book for Science*

Teaching has been considerably enlarged. This new, revised edition includes an expanded section in the biological sciences and a great deal of new material in earth and space sciences. The physical science section also contains considerably more chemistry teaching material than hitherto.

If science is to be perceived by pupils as a unity, key concepts which provide the foundation of many scientific disciplines need to be emphasized in its teaching. Basic scientific concepts such as matter and energy and their interrelation, and levels of organization in living things, provide the key themes for major chapters of the book.

While the style of the source book has been kept as recognizable as possible to its many users throughout the world, some major changes have been made in format and presentation, with the intention of making it easier to use, and an index has been included. The material has been grouped under four major chapter headings: 'Resources, Facilities and Techniques for Science Teaching', 'Physical Sciences', 'Biological Sciences' and 'Earth and Space Sciences'. Such headings are not meant to imply the compartmentalization of the teaching of science into these particular content areas. However, since the users of the book will be following widely differing curricula as far as content and organization of material is concerned, such an arrangement has been chosen to facilitate ease of reference, and no greater over-all integration of material has been attempted. An effort has been made to improve the accuracy of all the information presented and to include only experiments and equipment of proven reliability under various climatic conditions. Many figures and diagrams retained from the previous edition have been redrawn. Much attention has been paid to laboratory safety, including a new section on this subject. International and metric units are used throughout the book.

Users of the new source book are invited to submit to Unesco comments, criticisms and suggestions which may be incorporated in future editions.

Chapter One

Resources, facilities and techniques for science teaching

Some suggestions about science teaching

Possible resources in a rural area (ecological activities)

An *abandoned farm field* offers an excellent opportunity to observe the process known as succession. The earliest plants that seed into the field are called pioneer plants. As the field community (ecosystem) changes with time, some populations are replaced by others. This replacement of populations is called ecological succession. It is often possible to observe a mature area such as a forest adjacent to a recently abandoned field. It is interesting to study the various stages of development and to infer what the intermediate stages must be.

A *wood or forest* near the school may be instructive for discovering seasonal changes in animals and plants; studying habits of plants and animals; finding out where animals live; seeing how animal and plant life depend on each other; seeing how physical surroundings, such as moisture, temperature and amount of sunlight affect living things; finding examples of useful and harmful animals and plants. Possible use: arrange a field trip to observe and collect materials; bring selected materials into the classroom.

A *building under construction* may provide an opportunity to see how electrical wiring is installed; how a building is insulated; the different materials that are used; the difference between soil dug from the foundations and garden soil; how sewage is disposed of. Possible activities

include collecting examples of building materials for study—wires showing different kinds of electrical insulation, various kinds of heat-insulating materials, samples of soil, etc.; talking with workmen who are wiring the house, installing plumbing, or doing similar types of work; observing the procedure for locating and drilling a well if there is to be one; examining plumbing; if an out-door toilet is used, finding out where it is located in relation to the water supply and why this location was selected.

A *sawmill* may be instructive for learning how trees are selected for cutting; finding out how young timber is protected; learning which kinds of trees are considered most valuable and why; observing the use of machines; learning how lumber is made and cured; observing changes in animal and plant life when an area has been cleared. Possible activities include visiting the sawmill to observe the procedures; bringing back samples of wood to see growth rings; walking through woods to observe how trees are being cut; examining various machines to observe how they help workmen.

A *farm* may be instructive for observing various ways of preserving and storing food; caring for animals; growing garden vegetables and flowers; observing the use of machines in house, field, barn, garden, orchard; observing how buildings and grounds are protected against fire and how accidents are prevented.

A *vegetable and flower garden* may be instructive for studying how plants get enough light, moisture and other essentials for growth; learning how ground is prepared for planting, how plants are transplanted, and how seeds are dispersed; studying how flowers are self- and cross-pollinated and how seeds sprout and grow; learning what kinds of soil are suitable for the growth of different kinds of plants and how the soil is tested; observing how plants store food and how plants change with the seasons. Possible activities include visiting the garden to observe plants and methods of growth; making collections of seeds and fruits that show methods of dispersal; sprouting seeds in the classroom to learn more about how plants grow; performing experiments with plants to see the effects of light, temperature and moisture on growth; planting a school garden (if practicable) to learn more about how plants grow.

An *apiary* may be instructive for observing how bees are cared for; learning how hives are constructed and how they are prepared for cold weather; learning what happens when bees swarm and how they are handled safely and how bees are helpful to man; observing bees at work and learning how life goes on inside a hive; seeing an example of social insects and of insects that are useful to man.

A *creek or pond* may be instructive for observing kinds of plant life and the adaptations of stems, roots, leaves, flowers and fruit to a moist environment; learning how animals are adapted for life in or near water and contrasting this with land animals; observing how these animals and plants change as the seasons change; observing the food-getting and home-building habits of the animal life.

Using the resources

The value of the resources depends on how skilfully they are used. Each should be used for a definite purpose or purposes: to help solve a problem, to make a scientific principle more graphic, to increase the tendency of pupils to inquire about their environment.

In preparing for a trip, the teacher and children should have clearly in mind a definitely stated problem or problems. The teacher and perhaps a small committee of pupils should first go to the place to be visited by the class, to determine its suitability and accessibility.

Whenever the pupils plan to seek information from someone in the community, make sure that the informant understands the purpose of the visit, and keeps explanations easy enough for the pupils to understand.

Follow-up discussions should be carefully planned. Appropriate data should be used in solving the problem, and written records made of the findings whenever it seems likely that the children will have a use for the records.

Facilities for teaching science

Making a science corner in the classroom. Set aside a corner in the classroom and call it the Science Corner. If possible, obtain one or two tables which may be used for experimenting and display. Perhaps the school custodian will help you build shelves underneath the table for storage of materials, supplies and equipment, as described elsewhere in the book. Encourage the pupils to bring in materials to display in the Science Corner. These materials should never be allowed to remain on the table so long that their interest value is lost: the Science Corner should be a place of activity and change.

A science bulletin board. If children are encouraged, they will constantly bring to school interesting items they have clipped from newspapers or magazines. The science bulletin board provides a place to display such materials, as well as drawings and other things prepared in science classes. A good place for the science bulletin board is just above the tables in the Science Corner. The bulletin board can be made from softwood or from pressed fibreboard.

A museum shelf. Once children become interested, they are insatiable collectors. Some of the things they collect are bound to find their way to school. Such activities should be encouraged. One way to do this is to provide a museum shelf where collections or individual science items may be displayed.

Aquaria and terraria. Aquaria and terraria are a source of constant interest and provide a place where many important science phenomena may be observed. Directions for making and caring for aquaria are found in Chapter Three.

Cages for animals. Several types of animals can be kept in the classroom for observation. Some animals adjust to being caged better than others. Children may be encouraged to bring their pets to school for short periods of observation and study. Suggestions for building cages for animals will be found in Chapter Three.

Setting up a weather station. Simple weather instruments are described in Chapter Four. These can be made from materials available almost anywhere. Observing the weather changes from day to day is a source of interest and can form the basis for useful science lessons.

Growing plants. Small flowerpots placed along a window-sill where there is plenty of light will provide ample space for growing seeds and small plants. If more space is desired for some experiments, shallow wooden boxes may be obtained or made from new or scrap lumber.

Tropical conditions. In the tropics there are many causes of trouble in a laboratory, especially during the wet season. Materials perish, papers stick together, instruments rust, specimens go mouldy, lenses develop a fungus which quickly renders them useless and ruins their accurately ground surfaces. In addition, ants, termites and other insects continue their endless work of destruction.

Whatever can be kept in an air-tight container should be so kept. Glass jars, such as specimen jars with lids well greased, are ideal. Screw-capped bottles, e.g. those which have contained sweets,

are very useful. Metal containers, such as biscuit tins, cake tins, etc., can be rendered fairly airtight by strapping the joint between the lid and container with insulating tape.

Lenses of microscopes should be kept in a desiccator when not in use. A piece of string soaked in creosote and placed in the lens container with the eyepiece has been found useful for retarding the growth of mould.

During the rainy season, microscopes, galvanometers and other sensitive instruments should, if possible, be stored in an air-tight cupboard in which a 50-watt electric bulb is kept burning continuously. Needles can be inserted in a piece of cloth into which some vaseline has been rubbed. Metal instruments such as screw gauges, vernier callipers, tuning forks, etc., should be greased. The screws of retort stand bases, rings and clamps should be oiled frequently. Scalpels should be smeared with vaseline and kept in a case. The metal parts of tools should be rubbed over with an oily rag.

Laboratory safety

Activities and experiments are tremendous assets to a science lesson. Doing things can be fun; it can also be dangerous, and the science teacher must ensure that work is done in such a way that no accidents occur.

Many of our daily activities are potentially dangerous. Lighting a fire, crossing a street, driving an automobile or even taking a bath may result in an accident, but we don't stop doing any of these things just because of the danger involved. Rather, we teach our children the dangers involved in each of these activities so that they are able to enjoy the benefits of the activity and avoid the potential dangers. The same philosophy should be used in science lessons. Pupils should be taught the dangers inherent in each activity and how to avoid hazards. A number of precautions are outlined below.

Burns and fire. Perhaps the most common type of accident in the laboratory is a burn. Most burns can be avoided if pupils will remember, 'When you heat something, it gets hot—and may stay hot for some time'. Although obvious, this advice is commonly disregarded. Pupils will place a piece of glass or metal in a flame for several seconds, remove it and touch the end to see if it is hot. It is! Unfortunately, most hot objects cannot be distinguished from cold ones by sight, smell or sound. Only the sense of touch provides a test, and if the object is *too* hot, even a cautious touch can result in a burn. Another simple admonition should be given regarding fire. 'Don't put things that burn near open flames.' Pupils should realize that clothes, hair, paper, wood and *many* common chemicals catch fire rather easily. Burners that are not being used should be extinguished. An additional caution concerns alcohol burners: if a large amount of heat is reflected down on to the burner base (e.g. when heating inside a shiny metal can which serves as a windshield) the alcohol may vaporize inside the bottle and cause the mild burner to become a very hot blow torch.

Cuts and broken glass. Minor cuts are a second type of accident which is commonly encountered. There are generally three causes; broken glass tubing, falling glass containers, and gas generator explosions.

Virtually all cuts from glass tubing will be avoided if the tubing is wrapped in a towel before it is inserted into a stopper. The glass should be lubricated with glycerine or water, held with the towel and twisted as it is inserted into the stopper.

The ends of all glass tubing should be fire polished, but care should be taken to avoid sealing the end completely. If the end is accidentally sealed and the tubing is used as part of a gas-delivery system, the gas cannot get through the tube and an explosion may result. Always check gas-generator systems to be sure that passages are open. Special care must be taken when oxygen or hydrogen are being generated because of the additional explosive hazard. Procedures for the preparation of these gases which require the use of heat are *not* recommended.

It is obvious that the danger of cuts from fall-

ing glass objects will be greatly reduced if glass containers are stored on the floor, or on a shelf where there is little danger of them being accidentally knocked off. This is particularly true in the case of large stock bottles of acids, gases, or inflammable liquids.

When glass is broken, it should be discarded in a specially marked container rather than in normal waste baskets. Be considerate of the person who takes away the rubbish.

Heating substances in test tubes. When heating substances in test tubes, the tube should be moved back and forth across the flame and the mouth of the tube should be pointed away from nearby persons (see figure).

Heating substances in test-tubes

Heat source

Test tubes should never be filled to more than one-third to one-half capacity as a precaution against boiling over. When transferring materials from one container to another, hold the containers at arm's length.

Smelling and tasting. The nose is a delicate instrument which deserves protection. Be careful when smelling chemicals. The correct technique is to fan the gas towards the nose and sniff cautiously (see figure).

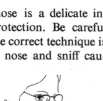

Smelling and testing

If you detect no odour, you may move closer and try again. The best rule for tasting is *'don't'*. Only substances that you know are absolutely harmless (such as pure table salt or sugar) should be placed in the mouth. Some chemicals are so toxic that a fraction of a gramme can be fatal.

Dangerous chemicals. Any chemical is potentially dangerous and should be treated as such. *Under no circumstances should pupils be allowed to perform unauthorized experiments and no experiments should be authorized unless you* know *that they are safe.* You may assume that the experiments described in this book are safe unless a caution is given. The potential danger should be clear from the note. If it is not, don't do the experiment. Just as it is the 'unloaded' gun that kills, it is the 'safe' experiment that often results in accidents. Substances such as sugar, sulphur, and powdered zinc are perfectly safe—normally. However, when mixed with good oxidizing agents such as chlorates or permanganates, they form explosive mixtures. A few of the more dangerous classes of chemicals are listed below with descriptions of their dangers.

Acids and bases. All 'strong' or 'mineral' acids such as hydrochloric, sulphuric or nitric acids are hazardous when concentrated. When diluted, they are relatively safe to handle and any spilled acid can be washed off with water. The greatest danger, of course, arises when acid comes into contact with the eyes. Safety glasses should be worn to protect the eyes from this danger. Some acids (e.g. sulphuric and nitric) are more hazardous because they are also good oxidizing agents. Organic acids are generally not so dangerous as mineral acids, but there are exceptions. Phenol (carbolic acid) and oxalic acid are hazardous, not because of their acidic properties but because they are toxic. Strong bases such as sodium hydroxide (caustic soda) and potassium hydroxide (caustic potash) can cause burns just like strong acids. Weaker bases such as calcium hydroxide (lime and water) can also cause burns if they are in contact with the skin for a long time. Dilute solutions of bases are relatively safe, but even dilute solutions should be quickly washed from the skin with plenty of water.

Oxidizing materials. (Oxidizing materials are chemicals that will support combustion or burning.) If they are placed in contact with materials that act as fuels (any organic matter) there is danger of an explosion or fire. Some of the more dangerous chemicals in this category are chlorates, peroxides, perchlorates and perchloric acid. Since sodium chlorate and potassium chlorate are fairly common chemicals, special note should be made of their dangers. These compounds are stable and may be handled safely providing proper caution is exercised. They should be kept away from strong acids since they react to produce toxic chlorine dioxide and may explode. They should be kept away from any substances that are easily oxidized such as suphur, sulphides, phosphorus, sugar, alcohols, organic solvents, ammonium compounds, powdered metals, oils or greases and dust of any kind.

Good habits. A few notes on practices which should be followed habitually are given below.
1. Always wear protective glasses when there is a danger of hot or caustic materials being splashed into the eyes.
2. Always read the label on reagent bottles *twice* and read it carefully. There is a very great difference between potassium chloride and potassium chlorate, between Mercury(I) chloride and Mercury(II) chloride, between manganese and magnesium.
3. Test tubes or any pieces of equipment which may expel a gas or liquid should be pointed away from all persons present.
4. Always check glassware for cracks prior to use.
5. Glasswear of all types should be placed at the back of the laboratory bench to prevent unnecessary breakage. Glass storage bottles should be placed at or near floor level.
6. All injuries, regardless of how minor, should be given medical attention immediately.
7. When diluting acids, the acid should be added slowly to the water not vice versa.
8. Good housekeeping is imperative in the labo-

ratory. Broken glass or scraps of metals and unused chemicals should be disposed of in appropriate containers. When chemicals are disposed of through the drain, always flush with plenty of water.

Using mercury. Surprising as it may be, mercury vaporizes even at the temperature at which water freezes. It produces an odourless, tasteless, and colourless vapour, the concentration of which depends upon the temperature. This vapour is toxic and can result in damage to the nervous system. Mercury enters the body readily by inhalation, by ingestion or through the skin. Chronic exposure may result in cumulative poisoning, indicated by nervous and psychic symptoms.

Protection against spillage. Mercury seeps into crevices, mixes with dust and penetrates into substances such as wood, tile, iron pipe and firebrick. Where mercury is used, floors should be smooth and impervious and cracks should be sealed by varnishing the floor surface. Floor spills must be cleaned up at once. If a spill occurs, the room must be evacuated and the windows opened to increase ventilation. Doors opening into hallways must be closed. Spilled mercury should be picked up immediately by water pump suction or by a wet sweeping compound. If these items are not available, a dustpan and broom, or a squeegee should be used to gather the mercury into one small mass. It should then be deposited into a tough plastic, glass, or metal container which can be sealed tightly. Contamination by numerous small globules left in cracks and crevices is still possible after the major portion of mercury has been removed. Calcium, sodium poly-

sulphide, or flowers of sulphur should be carefully administered to the contaminated area (the commercial agricultural dormant spray which contains sulphur is satisfactory for this purpose). These substances react with mercury to form an inert material which does not vaporize. It is recommended that a supply of flowers of sulphur be kept near by whenever mercury is handled.

Protection against skin contact. Every effort should be made to prevent skin contact with mercury liquid or vapour. Non-porous gloves and rubber-soled shoes should be worn, since leather will absorb mercury. One should wash hands thoroughly after contact with mercury to reduce absorption through the skin. Personal clothing should be checked for mercury following a spill, since mercury can be accidentally deposited in trouser turn-ups, pockets and crevices in clothing.

Storage. Mercury should be stored in a well-ventilated area where the containers are kept cool and protected from the direct rays of the sun. Storage on wooden floors is not desirable. Heavy linoleum, non-porous concrete and tile or enamel surfaces are all suitable, provided that joints and cracks are sealed and a smooth surface is achieved. Do not work with mercury, or store containers, near sources of heat or ammonia. Keep the seal on the mercury container closed tightly except when in use. This is important because a slow current of air passing over an open vessel of mercury at room temperature may pick up several milligrammes of mercury per cubic metre of air. If available, a leak-proof cabinet will provide an adequate storage space.

Useful tools and techniques

Tools

Clever experimenters can achieve a very high standard of workmanship with a surprisingly small number of tools. It is impossible to say that certain tools form a minimum amount of equip-

ment. In any case, the experimenter will probably acquire his equipment stage by stage and need not be deterred from starting a project because of some particular lack. As a beginning aim at acquiring the following:

Metalworking tools: vice, hack-saw, hammer,

screwdrivers, pliers (flat and round nose), cutting nippers, electric or other soldering iron, electric drill, twistdrills, taps and dies, various files, scriber.

Common woodworking tools: chisels, common saw, rip saw, steel plane, modern rasp or rasp plane, brace and bits, a variety of glues and cements, paints in various colours.

Glass cutting

1.1 *Making a straight cut*
A glasscutter does not *cut* glass, but splits it with a tiny wheel. If the wheel is sharp and it is drawn over the glass at the right speed and pressure, it makes a fine score or groove by slightly crushing or pulverizing the glass. The bevelled sides of the wheel act as wedges which push against the sides of the groove and pry the glass apart so that a crack is started. If a crack fails to start, tap the scratch or score with the ball end of the glasscutter. Before trying to make a finished cut, practise on a scrap piece to learn the speed and pressure required to obtain a smooth edge (see figure). Ordinary window glass comes in two thicknesses, single light and double light. Single light is thinner and easier to cut. Plate glass up to 0.6 cm in thickness can be cut in the same manner as ordinary window glass. Safety glass, which consists of two or more glass sheets cemented together by a transparent plastic, requires special cutting equipment.

1.2 *Cutting glass tubing*
One way to cut glass tubing is to score the surface with one forward stroke of a three-cornered file. The corner of a mill file will also work well. Make the score mark at right angles to the centre line of the tube so that the tube will snap squarely across. To snap the tube, place it on the bench top with a matchstick or toothpick directly beneath the upward-facing score mark. Then, holding one end securely, press down on the other end and the snap will be immediate. Another method frequently used is to scratch the glass tubing with a quick smooth file stroke,

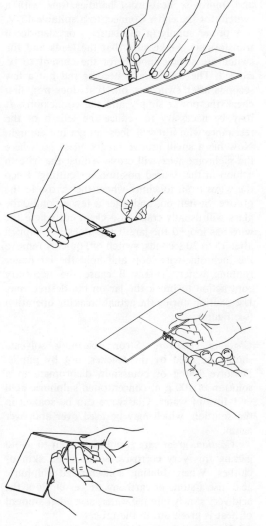

1.1 Glass cutting

then hold the scratched tubing firmly in both hands with one's thumbs pointing towards each other (but on opposite sides of the scratch) and snapping the glass tubing away from one's body. Fire-polish the cut ends.

1.3 *Resistance wire glass cutter*

Obtain about 60 cm of 24-gauge nichrome wire, and improvise heat-proof handles (one with a switch) for the ends. Connect to a suitable 12-V, 5-A power supply (a car battery, or step-down transformer). Make sure that the leads and the switch are strong enough for the current to be carried. The wire should become red hot a few seconds after switching on. If it does not, first check the power supply and the connexions. It may be necessary to reduce the length of the resistance wire if it still does not get hot enough. Now file a small groove on the glass jar where the nichrome wire will cross. Adjust the wire in a loop in the desired position for cutting. Keep the wires from touching where they cross in the groove. Switch on, and after a few seconds, the glass will usually crack in a clean cut where the wire has looped the jar. If this does not happen after 15 to 20 seconds, switch off, quickly remove the nichrome-wire loop and hold the jar under running water. This will cause the necessary contraction to break the jar on the desired line. Use caution during the actual breaking operation (see figure).

Cleaning glassware. Strong cleaning solvents should be used by the teacher, not by pupils. Dissolve 100 g of potassium dichromate in a solution of 100 g of concentrated sulphuric acid in 1 litre of water. Glassware can be soaked in the solution, which may be used over and over again.

Caution: great care should be taken to avoid getting this very corrosive solution on skin or clothes. When diluting concentrated sulphuric acid, use a stone or earthenware vessel. Pour the acid very slowly into the water, as a great amount of heat is given out in the process.

The teacher should use his knowledge of chemistry to remove stains of known origin. If dirty vessels have contained alkalis, or salts with alkaline reactions, then obviously the cleaning effect of a little dilute acid should be tried first; if the stain is due to potassium permanganate, then the effect of sodium sulphite solution, acidified with dilute sulphuric acid, should be tried,

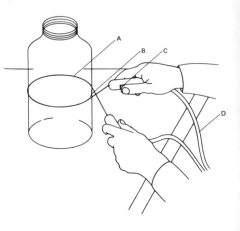

Resistance wire glass cutter
A 24 gauge nichrome wire
B small notch filed in side of bottle
C switch in handle
D connecting cable to supply source

etc. Alkalis slowly attack glass, and bottles which have contained caustic soda, etc., for a long time will never recover their original transparency.

Soldering

Soldering is used to join metallic surfaces such as copper, iron, nickel, lead, tin, zinc and aluminium. It is particularly useful for making electrical connexions, joining sheet metal and sealing seams against leakage. Electric soldering irons or guns are widely used for electrical connexions, but soldering may also be done with coppers which do not have an electrical heating element.

1.4 *Solders*

Most soft solders are alloys of tin and lead. Solders used for joining aluminium are usually alloys of tin and zinc or of tin and cadmium. The melting points of most tin-lead solders range from about 165° C upwards. Tin-lead solders are usually identified by numbers which indicate the respective proportions of tin and lead. The first number gives the percentage of tin, the second the percentage of lead. Solders with a high tin content are more expensive than those containing much lead. In general, solders which contain a high percentage of tin have lower melting points than those with a high percentage of lead; the former are best for electrical joints, whereas the latter have greater mechanical strength.

Solders are available in various forms, including bars, wires, ingots and powders. Wire solder is available with or without a flux core.

1.5 *Fluxes*

To make a good joint, the metal to be jointed, the tip of the soldering iron and the solder itself must be freed of dirt, grease, oxides and other foreign matter which would prevent the solder from adhering to the metal. Fluxes are used to clean the joint area, to remove the oxide film which is normally present on metal, and to prevent further oxidation. Fluxes also decrease the surface tension of the solder and thus make the solder a better wetting agent. Use a flux which is suitable for the metal to be joined, as shown below.

Metals	Fluxes
Brass, copper, tin	Rosin
Lead	Tallow, rosin
Iron, steel	Borax, sal ammoniac
Galvanized iron	Zinc chloride
Zinc	Zinc chloride
Aluminium	Stearine, special flux

Fluxes are generally classified as corrosive, mildly corrosive and non-corrosive. Non-corrosive fluxes are used for soldering electrical connexions and for other work which must be completely protected from any trace of corrosive residue. Rosin is the most commonly used non-corrosive flux. In the solid state, rosin is inactive and non-corrosive. When it is heated, it becomes sufficiently active to reduce the oxides on the hot metal and thus perform the fluxing action. Rosin may be obtained in the form of powder, paste, or liquid.

Rosin fluxes frequently leave a brown stain on the soldered metal. This stain is difficult to remove, but it can be prevented to some extent by adding a small amount of turpentine to the rosin. Glycerine is sometimes added to the rosin to make the flux more effective.

1.6 *Methods of soldering*

The following general considerations apply to most soldering work:

1. Be sure that all surfaces to be soldered are clean and free of oxide, dirt, grease or other foreign matter. If possible, the materials should be joined mechanically, so that the solder simply fixes the joint in place, just like carpenter's glue holds a woodworking joint.
2. Use a solder and flux which are appropriate for the particular job. Remember that the melting point of the flux must be *below* the melting point of the particular type of solder you are going to use.
3. Heat the surfaces just enough to melt the solder. Solder will not stick to unheated surfaces. However, you should be very careful not to overheat solder, soldering coppers or surfaces to be joined. In general, solder should not be heated much above the working temperature. As the temperature of molten solder is increased, the rate of oxidation is increased. When molten solder is overheated in air, more tin than lead is lost by oxidation.

1.7 *Electrical connexions*

To solder electrical connexions, use rosin-core solder. The reason for this is that it is usually difficult or impossible to wash off acid flux, whether applied in core solder or with a brush, from electrical gear. Any acid that remains from

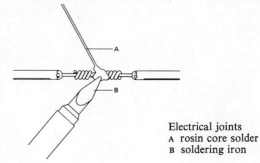

Electrical joints
A rosin core solder
B soldering iron

the soldering operation causes corrosion which cannot be tolerated. To solder electrical connexions, hold the soldering iron (copper) beneath the splice being soldered, with the greatest possible area of mechanical contact so as to permit maximum heat transfer. Apply the rosin-core solder to the splice (see figure). Be careful not to overheat electrical components.

1.8 *Torch soldering*
Torch soldering is often used for small jobs or for work which is relatively hard to reach. A propane torch or an alcohol torch may be used. The general procedure is to play the flame from the torch on to the surfaces to be joined and then apply cold solder in bar or wire form. The heated surfaces will melt the solder. As the solder melts, any excess solder should be wiped off with a damp cloth before it hardens.

Blueprints and diazo prints

Obtain two glass plates approximately 25 cm by 35 cm. Tape the edges and hinge the plates along one long edge.

1.9 *Blueprints*
Place a sheet of blueprint paper, greenish side up, on one of the two glass panes; lay the object to be printed on the paper (see figure); a photo negative, leaf or piece of lace will work well. Hold these in place with the second pane of glass.

Expose them to the sun for a period of 20 seconds to several minutes, depending on the brightness of the sky. Then wash the blueprint paper by placing it in a pan of water for several minutes. This removes all the remaining light-sensitive substance. Lay it on a smooth, flat table to dry.

1.10 *Diazo prints*
The procedure is the same as for blueprint paper up to the step of washing. Do not wash the paper in water. Instead, expose it to ammonia fumes in a large jar for several minutes. Then light can make no further changes on the paper (see figure).

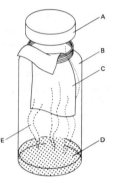

A lid
B wide mouth jar
C diazo paper
D household ammonia
E ammonia fumes

These experiences may induce some children to work with the more sensitive photographic materials used with cameras. Refer to local photo supply stores for more information about developing photographic films and prints.

1.11 *Making blueprint paper*
Prepare solutions of: potassium ferricyanide (10 g, 50 cm^3 of water) and ferric ammonium citrate (10 g, 50 cm^3 of water). The solutions are prepared separately and kept in a dark room or in subdued light. For use, mix equal quantities in subdued light and place in a shallow glass or enamelled tray. The paper is sensitized by brushing the mixed solution over it with a soft, wide brush or the paper may be placed on the surface of the solution and allowed to float there for a few seconds. After sensitizing, the paper should be hung to dry in the dark room.

Making some general equipment

Weighing devices

1.12 *A simple balance*
Punch four holes in an old tin can with a nail, spacing them equally round the circumference. Pass pieces of string through these holes and tie them together. Now attach this scale pan to a rubber band hung from a nail (see figure). If weights are not available, it is possible to graduate the balance using known volumes of water poured from a measuring jar and by making marks on the supporting stick opposite the edge of the pan. Stones can then be found which will give the same extension and these should be marked for future use as weights. Coins may also be used.

1.13 *A spring balance*
A coiled steel spring is protected from damage by enclosing it in a tube. The reading is made at the bottom of the tube on a graduated wooden plunger (see figure). First wind the spring, attach it by a screw eye to a piece of dowelling which will just fit into the tube selected (bamboo or plastic). Fasten the other end of the spring by a wire staple to a wooden stick which will slide in the tube. Fix the dowelling to the top of the tube and insert into it a hook for suspending the balance. Screw another hook into the wooden plunger which can now be graduated.

1.14 *Steelyards*
Either Roman or Danish steelyards can be improvised using short lengths of lead or iron water pipe as counter-weights and loops of wire as pivots (see figure).

The rod can be of either wood or metal; in the latter case notches can be filed on the underneath of the bar to indicate the balance points for various weights.

1.15 *A drinking-straw balance*
Obtain a small bolt which just fits inside the tube of a drinking straw, and screw it a few turns into one end. Determine roughly where this arrangement balances and punch a sewing needle through the straw to serve as a pivot. To ensure stability the hole should be made a little above the diameter of the straw (see figure overleaf).

Cut away the other end of the straw to form a small scoop. When the needle is in place set it across the edges of two microscope cover slips (or two razor blades) held parallel by a block of wood and a rubber band. Adjust the bolt until the straw balances at about 30 degrees to the horizontal. Support a piece of card vertically behind the scoop using a clothes peg or another piece of wood and a drawing pin; this will serve as a scale.

Hang a hair or a small piece of tissue paper from the scoop and notice the deflection. To

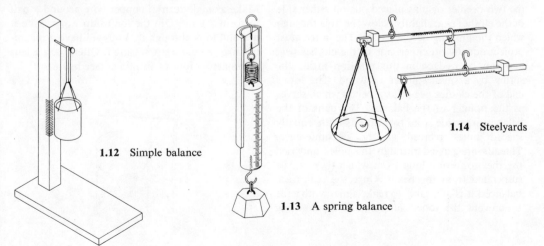

1.12   Simple balance

1.13   A spring balance

1.14   Steelyards

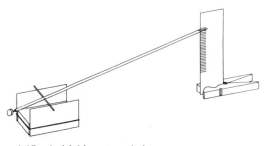

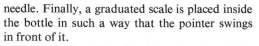

1.16   A sensitive beam balance

1.15   A drinking straw balance

obtain quantitative readings the scale must be calibrated. Aluminium foil from cigarette packets is suitable for making small weights.

Cut the foil into areas weighing 1 mg, 2 mg, etc., and place them in the scoop using a piece of copper wire bent to form tweezers. Record the positions of rest of the beam by making marks on the card. The sensitivity of the balance can be varied by adjusting the position of the bolt.

1.16   *A sensitive beam balance*
The materials needed for this balance include a clothes peg, a rigid knitting needle about 30 cm long, two pins or needles and a support such as a milk bottle or preserving jar (see figure).

The beam of the balance is made by passing the knitting needle through the hole in the spring of the clothes peg. The pivots for the beam are the two needles or pins placed one on either side of the clothes peg, slightly below the hole through which the knitting needle passes. The latter must project equally on either side of the clothes peg, and can be wedged in this position inside the spring by a small splinter of wood. The lower end of the clothes peg grips a pencil which serves as the pointer of the balance. The pans of the balance are made from two tin lids with equally spaced holes pierced at the circumference. Threads are passed through the holes and tied together to form a loop from which they can be suspended from the beam. Once the scales are balanced it is advisable to make a nick with a file to prevent the loops slipping off the knitting

needle. Finally, a graduated scale is placed inside the bottle in such a way that the pointer swings in front of it.

The weights may be coins, crown corks, matches, etc., correlated to standard weights. If none of the latter are available two similar small bottles may be used, one in each pan, and known amounts of water poured into one of them from some graduated vessel. Failing all else, an old syringe graduated in cubic centimetres may serve as a very small measuring cylinder. Fractional weights may be improvised by hanging a loop of wire from the beam.

Optical devices

1.17   *A simple magnifier*
Make a single turn of copper wire around a nail to form a loop. Dip the loop into water, take it out and look through it. You will have a magnifier like the earliest ones used. Often such a lens will magnify four or five times (see figure).

lens

If you tap the wire sharply against the edge of the glass a drop of water will fall off. Because of adhesion between the wire and the water, the liquid remaining will form a lens which is very thin at the centre, i.e. a concave lens.

1.18 *A water-drop magnifier*
Place a drop of water carefully on a plate of glass. Bring your eye close to the drop and look at something small through the water drop and glass. This serves as a simple magnifier.

1.19 *A model refracting telescope*
Arrange a long-focus lens on the end of an optical bench (see Chapter 2, experiment 2.219) pointing at some scene through a window. Bring a piece of white cardboard up on the opposite side of the lens to the place where the sharpest image of the scene is formed. Now bring a short-focus lens up behind the cardboard until the cardboard is a little nearer the lens than its focal length. Remove the cardboard and look through the two lenses at the scene.

1.20 *A projector for filmstrips or slides*
The base of the instrument is a piece of wood 40 by 10 by 3 cm. A plywood board 10 cm wide and 25 cm long fits into a groove cut across the base, and serves as filmstrip carrier. A hole 35 by 23 mm cut in this wood serves as an 'aperture' or gate to limit the light passing to one frame of the strip. The strip itself is held close to the gate, in a vertical position, by staples made from wire paper fasteners. These are easily bent to the width of the film; the ends are cut off short and sharpened with a file, and they can then be pressed into position on the plywood board. No reels are necessary. The strip can be moved on from one frame to the next by pulling on the end of the film; there is sufficient 'curl' to hold it stationary (see figure).

The lamp, which is an automobile headlamp in a holder mounted on a block, is adjustable; it can be slid between two strips of wood nailed to the base. A carafe or flask of water can be used as condensing lens and should be placed so that the whole of the gate is illuminated by the image of the lamp. When it has been so positioned, the

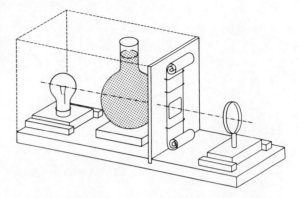

lamp and condensing flask are fixed in place with glue.

The object lens is mounted on a piece of wooden dowelling which is a fairly tight fit in a hole drilled into a block of wood arranged, like the lamp support, to slide between two wooden guides. The lens can then be adjusted by sliding the rod in or out of the hole so that the centre of the lamp, condenser and objective are all the same height above the baseboard.

A plywood, metal or cardboard case is required to enclose the lamp and the condenser as shown by the dotted line in the diagram. A darkened room is necessary for this apparatus. Commercial instruments using 100-watt bulbs can be used in a semi-darkened room, but the problem of dissipating the heat from the lamp is then considerable.

1.21 *A microprojector*
The optical system of this instrument is the same as that of the strip projector. The differences in construction are necessary because of the size of the objects (microscope slides or small objects similarly mounted) and the need to use a very short-focus objective to obtain high magnification. The lamp is a car headlight bulb, the condenser is a small glass bulb 1.5 to 2 cm in diameter blown on a piece of quill tubing, and the object lens is the objective of a commercial microscope.

The base of the apparatus is a small wooden trough 10 by 7 by 4 cm made by nailing two

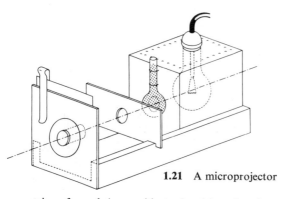

60 cm away from the end of the trough. Once the correct position for the slide has been found, the sawcuts are made in the edge of the trough and serve for all other slides used. This apparatus can also be used for projecting Newton's rings and diffraction phenomena (see diagram).

1.22 *Using a microscope as a microprojector*

1.21   A microprojector

If a very bright light source is used, the image from the eyepiece of a compound microscope can be reflected on to a screen with a mirror. A powerful slide projector is a good source of light.

strips of wood 4 cm wide to the sides of a piece measuring 10 by 5 by 1 cm. These sizes are not critical, and may be varied to suit the other materials. An end plate to support the objective (a piece of plywood 9 by 7 cm with a 2.5 cm hole in it) is fixed to the end of the trough.

A rectangular lamphouse fits into the channel. The lamphouse is easily improvised by fixing a car bulb and holder inside a household mustard or other rectangular tin. Holes drilled round the top and at the bottom provide ventilation, and a hole 1.5 cm in diameter serves to support the condenser. Copper wire passing round the stem of the bulb and through holes punched through the tin hold the condenser firmly in position.

The slide holding the object to be projected fits into grooves cut across the edges of the channel, and is thus held in a vertical plane so that the light from the condenser passes through it. The position of the grooves is determined in the way indicated below.

The microscope objective fits tightly into a hole in a piece of plywood, 7 by 4 cm, which is held in contact with the end plate by a trouser cycle clip and is adjusted so that the lens lies on the axis of the optical system.

The diagram shows the components mounted further apart than in actual practice; this is done to show the relative positions more clearly. In adjusting this apparatus the slide, lamphouse and condenser are moved forward together until the light passes through the objective and forms an image (of, say, a botanical specimen) on a ground glass screen 30 cm square placed about

Heat sources

1.23 *A candle burner*
A simple burner can be made using a tin lid or the bottom of a tin can and candles. The candles are fixed on to the lid by melted wax. The burner is most efficient if the length of the candles is kept approximately the same (see sketch).

1.24 *A tin can charcoal burner*
A large tin can at least 10 cm in diameter should be used. About halfway from the bottom mark off six triangular windows around the can, as shown in the first diagram. As shown in the second diagram cut along the sloping sides of each triangle to make the windows. Do not cut along the base line. Bend the triangular parts inward to form a shelf for the charcoal, file the sharp edges of the windows and make air holes.

1.25 *An alcohol lamp from an ink bottle*
Obtain an ink bottle with a metal top which screws on. Punch a hole in the centre of the metal top with a nail. Enlarge the hole until it is about 8 to 10 mm in diameter by turning a triangular file in it. Smooth the opening by using some hard, round object. Cut a piece of metal about 2.5 cm wide and 4 cm long from a soft metal can or piece of sheet metal. Roll this into a tube on a piece of dowel rod or other round wooden stick of suitable diameter to fit the opening in the top of the ink bottle. Insert the tube in the top and let it go about 1 cm into the bottle. The tube may be sol-

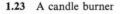

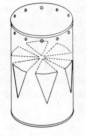

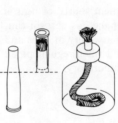

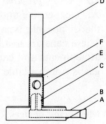

1.23 A candle burner   **1.24** A charcoal burner   **1.25** An alcohol lamp

1.26 A bunsen burner
A lead base
B brass gas supply tube
C brass jet tube
D barrel
E air hole to match air hole in barrel D
F ring of copper wire

dered to the top and along the seam. A wick may be made from cotton waste, a bit of cotton bath towel or from a bundle of strands of cotton string. Be sure to have enough wick to extend to the bottom of the bottle and cover it. Use denatured or wood alcohol.

In hot countries a cap should be made to cover the wick when the lamp is not in use. An old fountain-pen cap may serve the purpose. If a brass rifle cartridge is available it can be used to make both the tube and the cap by cutting it with a hacksaw at a suitable place (see figure).

1.26 *A bunsen burner*

If you do not already own a bunsen burner, it is a simple and instructive exercise to make one from scrap. There are no precise requirements for the size of the bunsen burner; the materials to make it can depend on what your scrap box contains. The brass tubes may be about 1 cm in diameter, but if you have brass tubing of different dimensions, go ahead and improvise.

Pieces of scrap lead are melted on a kitchen gas range in a strong tin box or discarded pot, then poured into a shoe-polish can. This casting (the can may be left on) forms a heavy base, A, for the burner (see figure). Holes are drilled vertically and horizontally as shown to take the brass tubes B and C. These must be slightly tapered and hammered into the lead. The gas supply tube B should extend about 2 cm into the base, but the jet tube C must just enter the horizontal hole. When the tube C has been tested for size, it is filled with a lead plug, cast in the tube. The lead is poured in around a centrally placed greasy sewing needle which, when extracted, leaves the jet hole. During the casting, the tube may be held in a shallow hole drilled in a wooden block with the needle precisely centred in the tube.

Brass tubes of suitable diameter are needed for the barrel D and the brass collar E in which matching air holes are to be cut. If the collar presents difficulty a tin tube may be substituted but will not look so well. If D is not already a good fit on the tube C it may be sealed in position with epoxy glue. The easiest way to make the matching air holes in D and E is first to run the tubes on to tapered dowel rods held in a vice. Flatten the tubes somewhat with a file, then use a 0.5-cm drill. Finally, trim the holes to shape with a round file, again using it to smooth the inner surface of the collar so that it turns easily on D. A ring of copper wire, F, is soldered on to D just above the collar, thereby preventing loss of the air adjuster.

With double air holes, i.e. drilled right through the tubes, it may be possible to overdo the air supply. In this case the burner may 'strike back' in use. It should not be allowed to burn for long at the lead jet or this may soften and close up the hole.

1.27 *Using propane gas*

An inexpensive and convenient source of heat is the disposable type of propane gas aerosol can. These cans are available in a variety of sizes and attachments and can be used when regular piped gas is not available in the laboratory.

Measuring devices

1.28 *A simple calorimeter*

Small soup tins can be found which fit loosely into a jam jar. If the top of the tin is cut off cleanly with a rotary type opener it serves as an excellent calorimeter (see figure).

The tin can be prevented from slipping into the jar either by a stout rubber band round the edge, or by cutting nicks in the rim and bending it slightly outwards. This form of suspension, and the low conductivity of glass and air contribute to its efficiency.

Expanded polystyrene (Styrofoam) drink cups are available in some countries and make excellent calorimeters. Other suitable calorimeters can be made using two metal cans or glass beakers. Select containers so that one will fit inside the other with at least 1 cm of space between them. Fill this space with glass wool or crumpled paper.

1.29 *A measuring jar or graduated cylinder*

Select several straight-sided glass jars of assorted sizes. Olive bottles are very useful for the making of graduated cylinders. Paste a strip of paper about 1 cm wide vertically on the outside of the bottle to within about a centimetre of the top.

Next, obtain a commercial graduated cylinder of about the same capacity as the bottle and pour enough water from it to fill the bottle nearly to the top of the paper scale. Draw a line across the paper scale and mark under it the number of cubic centimetres of water poured in. Repeat with lesser amounts of water to complete the scale.

Other useful suggestions

1.30 *A simple stand for heating*

A useful stand can be made by cutting away the sides of a tin can. It is convenient to make two or three of these to suit different burners and for use as stands. Holes should be punched along the upper edge to let the products of combustion escape (see figure).

1.31 *A heater*

A heater can be made from an old oil tin. Water is placed in the tin and heated from below. Iron wire is wrapped round a test tube and twisted to form a handle. The substance to be heated is placed in the test tube and the heater is used as shown in the diagram.

1.32 *Making distilled water*

A kettle can be used to provide steam, which is then condensed in a jam jar fitted with a large cork and immersed in a pan of cold water. Rubber tubing, adhesive tape or clay can be used to make the joint (see figure).

1.33 *An air oven*

A large tin can be used as an air oven. A hole through the lid fitted with a cork holds a thermometer, and the saucer or dish rests on a wire gauze bridge placed inside the tin (see figure).

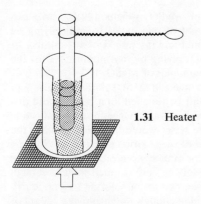

1.31   Heater

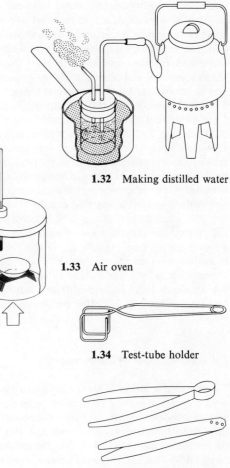

1.32   Making distilled water

1.34 *A test-tube holder*
A test-tube holder can be made by bending strong spring wire made of iron or brass into the shape shown in the diagram. Wire from a coat hanger works very well (see figure).

1.35 *Laboratory tweezers*
Very serviceable tweezers can be made from lengths of flexible strap iron used to put around boxes and crates for shipment (see figure).

    The tweezers shown are about 12 cm in length. One pair shown in the diagram can be made by brazing or riveting two pieces of strap iron together and then bending and cutting to the proper shape. The other pair shown were fashioned from a single 25 cm length of strap iron. The round head was made by pinning the centre of the strip around an iron rod of suitable diameter. The sides were then cut and shaped to size.

1.33   Air oven

1.34   Test-tube holder

1.35   Laboratory tweezers

Chemical solutions

Most of the chemical reactions that are studied in introductory science courses are reactions that occur in solution. Normally it is the material that is dissolved in the water (or other solvent) that is of interest. It is the dissolved material that is chemically changed. However, although the above is a good general rule, it is not always true. There are several activities in this book where water is changed chemically; that is, it takes part in the chemical reaction. Because we are normally interested in the material dissolved, it is important to know how much dissolved material (solute) is actually present in a particular volume of solution. For example, vinegar is a dilute solution of acetic acid. It is the acetic acid that gives the vinegar its sour taste and consequently the concentration is important. Most commercial vinegar is a 5 per cent solution; i.e. 100 g of vinegar contains 5 g of acetic acid. Percentage by weight is one way of describing the concentration of a solution. However, since chemical change involves interactions among *molecules*, it is convenient to express concentration in terms of molecules rather than weight. In other words, we want concentration defined in such a way that equal volumes of two different solutions of the same concentration will contain the same number of molecules. Such a concentration term is molarity. This is the concentration term used throughout this book and is abbreviated to 'M'. The notation that a solution is 1 M means that one litre of that solution will contain one mole (6.2×10^{23} molecules) of the solute.

Preparing solutions of known molarity

To prepare a solution of any given molarity you need only weigh out the number of moles needed and dissolve this weight of solute in enough distilled water to make one litre of solution. But how do you weigh out moles? For this, the molecular weight of the material must be known. An example will help.

Suppose you want to prepare a 2 M solution of $MgSO_4$. The first step is to determine the weight of 2 moles of $MgSO_4$. The weight of one mole is found by adding the atomic weights of *all* atoms represented by the formula and writing the total in grammes:

$$
\begin{aligned}
&\text{1 atom of Mg:} &1 \times 24.3 &= 24.3 \\
&\text{1 atom of S:} &1 \times 32.1 &= 32.1 \\
&\text{4 atoms of O:} &4 \times 16.0 &= \underline{64.0} \\
&&\text{Total} &\quad 120.4 \text{ (molecular} \\
&&&\qquad\quad \text{weight).}
\end{aligned}
$$

One mole of $MgSO_4$ weighs 120.4 g. We can find the weight of two moles by multiplying by two. (We could find the weight of any number of moles by multiplying by that number.) Thus the weight of two moles of $MgSO_4$ is 2×120.4 g $= 240.8$ g. To make a 2 M solution, 240.8 g of $MgSO_4$ would be weighed on a balance and dissolved in distilled water. Once the material is dissolved, more water would be added to make the total volume of the solution equal to one litre. A solution of the same concentration (2 M) could be made by dissolving half as much solid in enough water to make 500 ml of solution or one-fourth as much solid in enough water to make 250 ml of solution.

Some chemicals contain water of hydration (or crystallization); i.e. water is contained as part of the solid crystal. In these salts, the water must be considered part of the formula when calculating the weight of one mole of the solid. For example, magnesium chloride crystallizes as $MgCl_2$. $6 H_2O$. This means that six water molecules are included for each formula unit (molecule) of $MgCl_2$. The weight of one mole of $MgCl_2.6 H_2O$ is:

$$
\begin{aligned}
&\text{1 atom of Mg:} &1 \times 24.3 &= 24.3 \\
&\text{2 atoms of Cl:} &2 \times 35.4 &= 70.8 \\
&\text{12 atoms of H:} &12 \times 1.0 &= 12.0 \\
&\text{6 atoms of O:} &6 \times 16.0 &= \underline{96.0} \\
&&\text{Total} &\quad 203.1 \quad \text{(mole-} \\
&&&\qquad\quad \text{cular weight)}
\end{aligned}
$$

One mole of $MgCl_2.6 H_2O$ weighs 203.1 g.

The concentrations of most solutions used in this book need not be exact, and they can be prepared by weighing the solute to the nearest gramme.

Preparation of reagents

Volumes are stated in millilitres (ml) and litres (l). One millilitre is equivalent to one cubic centimetre (cm³ or cc) for all practical purposes. Masses are indicated in grammes (g). The relation to molar solution (M) is indicated in many cases. Distilled water should be used.

Laboratory reagents for general use

1.36 *Dilute acids, 3 molar*
Use the amount of concentrated acid indicated and dilute to one litre.

Acetic acid. 3N. Use 172 ml of 17.4 M acid (99 – 100 per cent).

Hydrochloric acid. 3N. Use 258 ml of 11.6 M acid (35 per cent HCl).

Nitric acid. 3N. Use 195 ml of 15.4 M acid (69 per cent HNO_3).

Sulphuric acid. 6N. Use 168 ml of 17.8 M acid (95 per cent H_2SO_4). (In this case, pour the sulphuric acid slowly into three-quarters of the final volume of water, and add water to make up to one litre when the solution has cooled.)

1.37 *Dilute bases*
Ammonium hydroxide, 3M. Dilute 200 ml of concentrated solution (14.8 M, 28 per cent NH_3) to 1 litre.

Calcium hydroxide, 0.02 M. Saturated solution, 1.5 g of Ca $(OH)_2$ per litre. Use some excess, filter off $CaCO_3$ and protect from CO_2 of the air.

Sodium hydroxide, 3M. Dissolve 126 g of the sticks (95 per cent) in water and dilute to 1 litre.

General reagents

1.38 *Accumulator electrolyte (lead accumulator)*
The relative density of the sulphuric acid in various conditions of the battery is: fully charged, 1.28; half charged, 1.21; discharged, 1.15. These figures are approximate. The recommendations of the makers for filling and initial charging, usually printed on the battery, should be followed.

A rough guide to the making of a solution of sulphuric acid of relative density 1.28 is as follows. Concentrated sulphuric acid is added slowly, with stirring, to a strong beaker two-thirds full of distilled water, until the solution is almost boiling. The solution is allowed to cool and more acid is added, with similar precautions, until the solution is again almost boiling. After cooling to room temperature, the relative density is adjusted by the addition of more acid or more water, according to the hydrometer reading. Always take great care when handling concentrated acid. Wear protective glasses and clothing.

1.39 *Aqua regia*
Mix 1 part concentrated nitric acid, HNO_3, with 3 parts of concentrated hydrochloric acid, HCl. This formula should include one volume of water if the aqua regia is to be stored for any length of time. Without water, objectionable quantities of chlorine and other gases are evolved.

1.40 *Bismuth chloride, 0.17 M*
Dissolve 53 g of bismuth chloride, $BiCl_3$, in 1 litre of dilute hydrochloric acid, HCl. Use 1 part concentrated HCl to 5 parts water.

1.41 *Bismuth nitrate, 0.083 M*
Dissolve 40 g of bismuth nitrate, $Bi(NO_3)_3 \cdot 5H_2O$, in 1 litre of dilute nitric acid, HNO_3. Use 1 part concentrated nitric acid to 5 parts water.

1.42 *Copper(II) sulphate, 0.5 M*
Dissolve 124.8 g of copper(II) sulphate, $CuSO_4 \cdot 5H_2O$, in water to which 5 ml of concentrated sulphuric acid, H_2SO_4, has been added. Dilute to 1 litre.

1.43 *Iron(III) chloride, 0.5 M*
Dissolve 135.2 g of iron(III) chloride, $FeCl_3 \cdot 6H_2O$, in water containing 20 ml of concentrated hydrochloric acid, HCl. Dilute to 1 litre.

1.44 *Iron(III) sulphate, 0.25 M*
Dissolve 140.5 g of iron(III) sulphate, $Fe_2(SO_4)_3 \cdot 9H_2O$, in water containing 100 ml of concentrated sulphuric acid, H_2SO_4. Dilute to 1 litre.

1.45 *Iron(II) ammonium sulphate, 0.5 M*
Dissolve 196 g of iron(II) ammonium sulphate, $Fe(NH_4SO_4)_2 \cdot 6H_2O$, in water containing 10 ml of concentrated sulphuric acid, H_2SO_4. Dilute to 1 litre. Prepare fresh solutions for best results.

1.46 *Iron(II) sulphate, 0.5 M*
Dissolve 139 g of iron(II) sulphate, $FeSO_4 . 7H_2O$, in water containing 10 ml of concentrated sulphuric acid, H_2SO_4. Dilute to 1 litre. This solution does not keep well.

1.47 *Lime water*
Lime is not very soluble in water, but the solution required for class use is easily made by adding 10 g of slaked lime to 1,000 ml of distilled water. After shaking, allow it to settle, and decant off the clear liquid.

1.48 *Litmus solution*
Powder the litmus and boil with water for five minutes. Filter the solution and store in a bottle. A fresh solution will have to be made from time to time.

1.49 *Mercury(I) nitrate*
Use 1 part mercury (I) nitrate, $Hg_2(NO_3)_2$, 20 parts water and 1 part concentrated nitric acid, HNO_3.

1.50 *Sea water*
A useful substitute for sea water can be obtained by dissolving the following in 2 litres of water:

    45.0 g sodium chloride
    3.5 g magnesium sulphate
    5.0 g magnesium chloride
    2.0 g potassium sulphate

1.51 *Sodium sulphide, 0.5 M*
Dissolve 120 g of sodium sulphide, $Na_2S . 9H_2O$, in water and dilute to 1 litre. Alternatively, saturate 500 ml of 1 M sodium hydroxide, NaOH (21 g of 95 per cent NaOH sticks), with H_2S, keeping the solution cool, and dilute with 500 ml of 1 M NaOH.

1.52 *Tin(II) chloride, 0.5 M*
Dissolve 113 g of tin(II) chloride, $SnCl_2 . 2H_2O$, in 170 ml of concentrated hydrochloric acid, HCl, using heat if necessary. Dilute with water to 1 litre. Add a few pieces of tin foil. Prepare a fresh solution at frequent intervals.

Special solutions and reagents

1.53 *Alloys*
Lower melting alloys. These may be produced by using a bunsen burner. Where both bismuth and lead occur together in an alloy, the bismuth and lead are melted together, and then the other ingredients added. The temperature should not be higher than necessary so as to prevent excess oxidation. The parts indicated are by weight.

Alloy	Lead	Tin	Bismuth	Cadmium
Wood's metal	4	2	7	1
Solder	1	1	0	0
Electric fuse alloy	8.5	2.5	1.3	0

Higher melting alloys. These may be produced in a furnace. The copper should be melted first, and the other metals added to it.

Alloy	Copper	Tin	Zinc
Bronze	80	5	15
Brass, malleable	58	0	42
Brass, casting	72	4	24

1.54 *Benedict's solution (qualitative reagent for glucose)*
With the aid of heat, dissolve 173 g of sodium citrate and 100 g of anhydrous sodium carbonate, Na_2CO_3, in 800 ml of water. Filter, if necessary, and dilute to 850 ml. Dissolve 17.3 g of copper sulphate, $CuSO_4 . 5H_2O$ in 100 ml water. Pour the latter solution, with constant stirring, into the carbonate-citrate solution, and make up to 1 litre.

1.55 *Brom thymol blue*
Dissolve 0.5 g of brom thymol blue in 500 ml of water. Add a drop of ammonium hydroxide to turn the solution deep blue in colour.

1.56 *Cements and waxes*

Cementing compounds can be made up easily by following the recipes below.

Cements

Acid-proof cement: 1 part rubber solution; 2 parts linseed oil; 3 parts powdered pipeclay.

Aquarium cement: (a) Equal parts of powdered sulphur, ammonium chloride, and iron filings are mixed. Boiled linseed oil is then added and all are mixed thoroughly. White lead is added to form a thick paste. The cement should be applied while fluid.

(b) Mix red lead with sufficient gold size to make a smooth paste and apply immediately. Allow a few days to set, and rinse the aquarium before using.

Celluloid cement: celluloid scraps can be dissolved in acetone or amyl acetate. This cement is useful when making up small accumulators.

Cement for iron: 90 parts fine iron filings, 1 part flowers of sulphur, 1 part ammonium chloride. Mix to a paste with water immediately before use.

Waxes

Chatterton's compound: 1 part archangel pitch, 1 part resin. Melt these together and add 3 parts of crêpe rubber in small pieces.

Faraday's cement: 5 parts resin, 1 part beeswax, 1 part yellow ochre. Melt the resin and wax together in a tin and stir in the ochre.

1.57 *Clarke's soap solution (for estimation of hardness in water)*

Dissolve 100 g of pure powdered castile soap in 1 litre of 80 per cent ethyl alcohol and allow to stand overnight (solution A).

Prepare a standard solution (B) of calcium chloride, $CaCl_2$, by dissolving 0.5 g of calcium carbonate, $CaCO_3$, in hydrochloric acid, HCl (relative density 1.19), neutralize with ammonium hydroxide, NH_4OH, and make slightly alkaline to litmus, and dilute to 500 cm³. One millilitre is equivalent to 1 mg of $CaCO_3$.

Titrate solution A against solution B (solution A in the burette). Dilute A with 80 per cent ethyl alcohol until 1 ml of the resulting solution

is equivalent to 1 ml of B, after making allowance for the lather factor (the amount of standard soap solution required to produce a permanent lather in 50 cm³ of distilled water). One cubic centimetre of the adjusted solution after subtracting the lather factor is equivalent to 1 mg of $CaCO_3$.

1.58 *Cupric oxide, ammoniacal: Schweitzer's reagent (dissolves cotton, linen and silk, but not wool)*

(a) Dissolve 5 g of cupric sulphate in 100 ml of boiling water, and add sodium hydroxide until precipitation is complete. Wash the precipitate well, and dissolve it in a minimum quantity of ammonium hydroxide.

(b) Bubble a slow stream of air through 300 ml of strong ammonium hydroxide containing 50 g of fine copper turnings. Continue for one hour.

1.59 *'Dead black'*

This is useful for painting the inside of 'light' apparatus, so that unwanted reflected light may be eliminated, rays made less diffused, and images made sharper. Lamp black is mixed with gold size, and turpentine added, with constant stirring, until the mixture is sufficiently thin for use as a paint.

1.60 *Dyeing*

The dyeing of cotton should be preceded by removing the sizing from the fabric. This is accomplished by boiling it for 5 minutes in a dilute solution of HCl (hydrochloric acid). This solution is made by adding 1 part of concentrated HCl to 10 parts of water. The following formula makes a satisfactory dye:

    Congo red 0.5 g

    $NaHCO_3$ (sodium bicarbonate) 2.0 g

    Na_2SO_4 (sodium sulphate) 1.0 g

    H_2O (distilled) 200.0 mil

The fabric should be boiled for 4 to 5 minutes and then rinsed in cold water and dried.

Instead of the congo red, methylene blue or primuline brown may be used. The dye and salts should be mixed together first and then added slowly, with stirring, to the water.

White silk, rayon or wool may be dyed in the

same way. Heat a piece of white cotton fabric for 10 minutes in a dilute solution of $(NH_4)_2SO_4$ (ammonium sulphate). It should stand for a few minutes in dilute NH_4OH (ammonium hydroxide), after which it is rinsed. White silk may be mordanted by boiling for 5 minutes in a tannic acid solution. It should then stand for a few minutes in a solution of tartar emetic. The effect of the mordant may be studied by boiling the mordanted and unmordanted pieces of cotton and silk in alizarin solution for a few minutes, after which they are rinsed and dried.

Boil samples of mordanted and unmordanted cotton and mordanted and unmordanted silk in a solution of malachite green (or methylene blue) for 5 minutes. They are then rinsed and dried. The malachite green solution is made by dissolving 1 g of dye in 200 g of water. Two hundred grammes of water are acidified with acetic acid. Forty grammes of the dye solution are added to the acidified water.

The development in the fibres of colours known as ingrain or developed dyes requires the use of three solutions. The first consists of 0.1 g of primuline and 0.1 g of $NaHCO_3$ (sodium bicarbonate) dissolved in 100 cm³ of water. Boil a strip of unsized cotton in this solution for 1 minute, then transfer it to the second solution. This solution is made by adding 0.5 g of $NaNO_2$ (sodium nitrite) and 3 cm³ of HCl to 100 cm³ of water. The strip is left in this bath for 15 minutes and is then transferred to the developing bath. The developing bath is made by dissolving 0.05 g of NaOH (sodium hydroxide) and 0.05 g of phenol in 100 cm³ of water. (Instead of phenol, alpha naphthol or resorcinol may be used.) The solution should be kept warm and the cloth allowed to remain in it for 20 minutes, after which it is rinsed and dried.

1.61 Electroplating solutions
1. *Copper.* About 100 g of copper sulphate crystals are dissolved in about 300 cm³ of water; 6 g of potassium bisulphate and 5 g of potassium cyanide are then added. The solution is made up to 450 cm³. (The solution should be kept cold while it is being made.)

2. *Silver.* About 20 g of sodium cyanide (poison) and 40 g of crystalline sodium carbonate are dissolved in about 500 cm³ of water. About 20 g of silver nitrate are dissolved separately in 250 cm³ of water. The second solution is added slowly to the first, and the mixture made up to 1 litre.

The current to be passed through the solutions depends on the area of the electrode upon which the metal is to be deposited. It should not exceed about 2 amps for 100 cm² of surface. About four to six volts direct current source is usually convenient, and may be obtained from a six-volt car battery. The current should be proportionately less if the electrode is smaller. The deposited metal will not present the expected bright and shining appearance until it has been burnished, by rubbing, for instance, with a bone spatula or some other hard, smooth, non-metallic object.

1.62 Fehling's solution (reagent for reducing sugars)
1. *Copper sulphate solution.* Dissolve 34.7 g of $CuSO_4\ 5H_2O$ in water and dilute to 500 cm³.
2. *Alkaline tartrate solution.* Dissolve 173 g of potassium sodium tartate (Rochelle salts, $KNa\ C_4H_4O_6.4H_2O$) and 50 g of NaOH in water and dilute when cold to 500 cm³.

Mix equal volumes of the two solutions at the time of using.

1.63 Fluorescein solution
This is useful because the track of a ray of light travelling through a dilute solution of fluorescein can be seen very clearly. One gramme of fluorescein is dissolved in 100 ml of industrial, or methylated, spirit.

1.64 Heat-sensitive paper
A solution of cobalt chloride in water is added to a solution of ammonium chloride in water. (The proportions do not matter.) The solution is diluted until it is pale pink. Filter paper soaked in the solution and allowed to dry appears to be almost colourless, but on heating it will turn a bright green colour.

1.65 *Iodine, tincture of*
To 50 cm^3 of water add 70 g of iodine I_2, and 50 g of potassium iodide, KI. Dilute to 1 litre with alcohol.

1.66 *Nessler's reagent (for ammonia)*
Dissolve 50 g of KI in the smallest possible quantity of cold water (50 cm^3). Add a saturated solution of mercuric chloride (about 22 g in 350 cm^3 of water will be needed) until an excess is indicated by the formation of a precipitate. Then add 200 cm^3 of 5 N sodium hydroxide NaOH and dilute to 1 litre. Allow to settle and draw off the clear liquid.

1.67 *Oxygen absorbent*
Dissolve 300 g of ammonium chloride in 1 litre of water and add 1 litre of concentrated ammonium hydroxide solution. Shake the solution thoroughly. For use as an oxygen absorbent, a bottle half full of copper turnings is filled nearly full with the NH_4Cl-NH_4OH solution and the gas passed through.

1.68 *Silvering solution (for depositing a bright silver mirror on glass)*
For *solution A*, 12.5 g of silver nitrate are dissolved in 100 cm^3 of water, and 32.5 g of sodium potassium tartrate are dissolved separately in 100 cm^3 of water. These two solutions are mixed, warmed to 55° C, and kept at that temperature for 5 minutes. The mixture is then cooled and the clear liquid poured off from the precipitate and made up to 200 cm^3.

For *solution B*, 1.5 g of silver nitrate are dissolved in 12 cm^3 of water. Dilute ammonium hydroxide solution is added until the precipitate first formed is almost entirely redissolved. The liquid is made up to 200 cm^3.

Solutions A and B are then mixed. (The surface to be silvered, after careful cleaning to free it from all traces of grease, should be suspended upside down in the solution, just below the surface. The solution can be put into a clean test-tube, or small flask and a mirror will be deposited on the inside of the vessel. The solution may be slightly warmed to hasten the deposition of the silver.)

1.69 *Sodium hydroxide (for CO$_2$ absorption)*
Dissolve 330 g of NaOH in water and dilute to 1 litre.

1.70 *Starch solution*
(a) Make a paste with 2 g of soluble starch and 0.01 g of mercuric iodide, HgI_2, with a small amount of water. Add the mixture slowly to 1 litre of boiling water and boil for a few minutes. Keep in a glass-stoppered bottle. If other than soluble starch is used, the solution will not clear on boiling; it should be allowed to stand and the clear liquid decanted.

(b) A solution of starch which keeps indefinitely is made as follows: mix 500 cm^3 of saturated NaCl solution (filtered), 80 cm^3 of glacial acetic acid, 20 cm^3 of water and 3 g of starch. Bring slowly to the boil and boil for 2 minutes.

1.71 *Tannic acid (reagent for albumen, alkaloids and gelatin)*
Dissolve 10 g of tannic acid in 10 cm^3 of alcohol and dilute with water to 100 cm^3.

Chapter Two

Physical sciences

Chemistry

Introduction

The experiments are set out in the simple, progressive development of concepts as outlined under the main headings. Teachers may select and successfully tackle any experiment that appeals, but on reflection they may feel the need to go back to some previous experiments. For example, if they attempt an experiment on electrical energy from chemical reactions, they may subsequently feel the need to investigate electrical conductivity and the properties of ions; or they may find there is need to know more about criteria for purity before trying to separate substances.

It is anticipated that the experiments will stimulate discussion not only on important ideas of chemistry but also on how a community applies and may benefit from these ideas. Industry has to separate and purify substances before it begins to change them to more useful substances. Industry uses filter plants, classifiers and cyclones, and takes advantage of differences in melting point, boiling point, solubilities and densities. It is envisaged that the experiments in these sections will cause pupils to question how industry tackles the problem on a much larger scale.

In their immediate surroundings pupils see that there are problems of selecting construction materials such as plasters, cements and concrete; problems of using metals and the soldering, alloying and conductivity of these metals.

A number of questions appear among the experiments. It is hoped that, as a result of their experiments, pupils will be stimulated to ask similar questions, discuss problems, and search out more information from books so that their understanding of chemistry may be enriched.

⬆ In all diagrams this symbol represents the source of heat.

The bunsen burner

2.1 *Investigating a bunsen burner*
When heating things we want to know which is the hottest part of the bunsen flame.

A. First close the air-hole and turn the gas-tap full on. Light the gas and hold a piece of wire in different parts of the flame, moving it from the bottom to the top. Where is the hottest point? Now open the air-hole. Again hold the wire in the flame, moving from the bottom to the top. Where is the hottest part in this flame? Pupils should be able to compare the two flames and say which has the hottest point.

B. Close the air-hole. Hold a test-tube with its bottom end just above the flame. Carbon may be deposited on the glass. Is it unburned carbon which gives the yellow colour to the flame? If powdered charcoal is sprinkled into the flame, does this give the same effect?

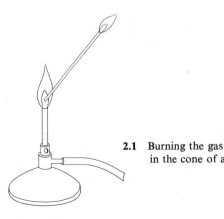

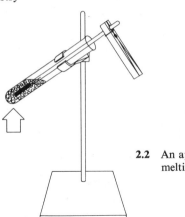

2.1 Burning the gas
in the cone of a flame

2.2 An approximate
melting point

C. Open the air-hole again. Is carbon deposited on a test-tube held in this flame? Air mixing with the gas helps it to burn more rapidly and efficiently. What is in the cooler inner cone? Hold a splint of wood in the flame so that it passes through the inner cone. Which part of the splint burns? Put a piece of tubing with one end in the inner cone as shown in the diagram. Gas comes out of the other end which can be ignited. Can a gas burner be designed to give a hotter flame?

D. Investigate a candle flame and the flame of a spirit lamp in a similar way. Which is the hottest part of the flame? Does the flame contain unburned carbon particles? Is there an inner cone of unburned gases?

The identification of pure substances

2.2 *Comparing the melting points of naphthalene and stearic acid*

Naphthalene (moth-balls) is a convenient substance to use. It has a melting point of 80.2° C. Put about 2 cm of naphthalene in a 100 by 16 mm test-tube clamped in a retort (see diagram). Hold a thermometer with its bulb in the naphthalene.

With a small flame gently heat the test-tube and watch the thermometer reading closely. What is the temperature of the naphthalene when it melts? Stop heating when the naphtha-

lene melts. Let it cool. What is the temperature at which the naphthalene solidifies again?

With a clean test-tube and thermometer repeat the experiment using stearic acid or any other substance with a melting point below 100° C if the thermometer reads to 110° C.

2.3 *A more accurate way of finding the melting point*

Put a very small amount of naphthalene in a capillary tube sealed at one end. (The capillary tube can be pulled out from glass tubing.) Attach the capillary tube, with sealed end down as shown in the diagram, to a thermometer by means of a rubber band and heat in a beaker of water on a tripod. The rubber band may be cut from a piece of rubber tubing. The thermometer may be used to stir the water, but make sure that no water enters the capillary tube. Slowly raise the temperature of the water and watch for the temperature at which the naphthalene melts. Record this temperature. Allow the tube to cool and record the temperature at which the naphthalene solidifies. Take the average of these two values. Which of the last two experiments appears to give the most accurate melting point? Repeat the last experiment using stearic acid.

2.4 *Impurities affect the melting point of a substance*

Mix a little stearic acid with the naphthalene, thus making the naphthalene impure. Look for changes in the melting point. Impurities lower the melting point.

2.5 *The boiling point of water*

A. Put a little water in a test-tube and hold a thermometer with the bulb just under the water as shown in the figure. Add a few granules, or boiling chips, to prevent bumping. Bring the water to the boil using a very small flame. Read the thermometer. Is there any change in the reading if the thermometer touches the bottom? What explanation would the pupil give?

B. Suggest that the pupils find out whether the boiling point of water is dependent on the amount of water present.

2.6 *The boiling point of inflammable liquids*

A. What other colourless liquids would the pupil know? Some of these liquids are very inflammable; for example, alcohol and acetone. A different way of heating these liquids must be used. First put the alcohol or acetone into the test-tube, about 2 cm deep, and put the thermometer into the liquid. Boil some water at a distance from the test-tube, then pour the hot water into the beaker so that the level is higher than the alcohol in the test-tube (see figure). Stir the alcohol gently with the thermometer and watch the reading of the thermometer. What is the boiling point of the alcohol? Can pupils explain what makes this a safe method of finding boiling points of inflammable liquids?

B. Another safe method using small amounts of inflammable liquids is the following. A piece of glass tubing about 8 cm in length and external diameter 2–3 cm is sealed at one end. A little of the liquid to be treated is put into this tube. A capillary tube, sealed at one end (like the one used for the melting point), is put into the liquid with the sealed end up and the open end in the liquid (see figure). The tube plus liquid and the capillary tube are then secured to the bulb of a thermometer by a rubber band. This is held in a beaker of water which can be heated gently by a bunsen flame. As the temperature is raised,

A

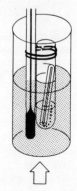

B

bubbles slowly come out of the capillary, but when the boiling point is reached the bubbles suddenly come out as a steady stream. Read the temperature. Then let the water cool and read the temperature again when the steady stream of bubbles ceases. The boiling point is the average of these two readings. Find out the boiling point of benzene by this method.

2.7 *The boiling point of a mixture of two liquids*
The boiling points of benzene and ethanol are not very different. Use the method described above to find out the boiling points of a few different mixtures of benzene and alcohol. Discuss with the pupils whether a pure substance could be identified by its melting point or boiling point.

2.8 *Pressure affects the boiling point*
Put some water in a test-tube with a side-arm or with an extra outlet through the cork. Put some granules in the water to stop bumping. Moisten the bung carrying the thermometer before fixing it in the tube (see figure). Heat the gauze strongly and boil the water. What is the temperature reading? Now stop heating, and connect a water pump to A. When the water stops boiling turn the water pump full on. Replace the bunsen and heat again. How has the pressure changed in the tube? At what temperature does the water boil now?

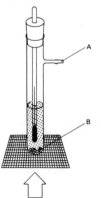

Decreasing the pressure on boiling water
A to water pump
B granules to stop bumping

Kenya is a country of East Africa. At Mombasa, on the coast, water boils at 100° C, but at Nairobi water boils below 95° C. Can you explain this?

2.9 *Comparing the different solubilities of substances in water*
A selection of salts taken from the laboratory shelves will demonstrate that each has a different solubility in water. Take 5 g samples and attempt to dissolve each in 15 cm³ of water contained in a test-tube. (Solubilities of salts will be given in data books as grammes dissolved in 100 g of water at a particular temperature, usually 20° C.) Each test-tube should be stoppered and shaken vigorously for the same period of time. This will demonstrate that solubility is one of the characteristics of a particular substance. The substances might include sugar, common salt, potassium nitrate, calcium sulphate, etc.

2.10 *Investigating the effect of temperature on solubility*
The solubility of potassium dichromate is approximately 5g/100 g in cold water and approximately 95 g/100 g in hot water. This is quite a large variation and may be demonstrated in the following way. Make up about 50 cm³ of a saturated solution of potassium dichromate at approximately 60° C. Pour the clear solution into a clean beaker and keep the temperature of this beaker at 40° C until crystals have ceased to form. Then pour the clear solution from this beaker into a third clean beaker. Avoid getting crystals into the beaker. Allow the solution to cool to room temperature. More crystals will form as it cools to room temperature. The experiment demonstrates that a saturated solution contains less dissolved solid at a low temperature than at a higher temperature.

2.11 *Finding out the solubility of a substance in water at a given temperature*
Put about 50 cm³ of water in a beaker and add baking powder (sodium bicarbonate) gradually whilst stirring. (Potassium sulphate is an alternative substance.) Stir until no more will dissolve,

i.e. until a saturated solution is obtained. Determine the temperature of the saturated solution. Weigh a clean evaporating dish. Pour some of the clear saturated solution into the evaporating dish and weigh again to give the mass of the solution. Carefully evaporate the solution to dryness. Weigh again to find the mass of the sodium bicarbonate dissolved. The mass of water is also found from these weighings. Thus the solubility of the baking powder can be calculated in g per 100 g water at a particular temperature.

2.12 Investigating the effect of particle size on solubility

Compare the rate of dissolving of coarse particles of salt with fine table salt, or large crystals of copper sulphate with crushed fine particles of copper sulphate. Add 4 g of coarse salt to one test-tube half filled with water, and 4 g of fine table salt to a second test-tube containing the same amount of water. Stir or shake both tubes equally and at the same time. Pause after every few seconds to observe the amount of undissolved salt left in each tube. The small particles dissolve faster than the large particles.

2.13 Investigating different solvent types

The solubility of common salt and iodine in the three solvents, water, alcohol and carbon(IV) chloride demonstrates this solvent effect. Fill three test-tubes one-third full; one with water, one with commercial methylated alcohol and the third with carbon(IV) chloride. To each add one spatula (about 1 g) of salt, then stopper and shake. It is found that the salt dissolves readily in the water, not so readily in the alcohol and very little in the carbon(IV) chloride. Prepare three more test-tubes with these three solvents. This time use a very small amount of iodine—just a few crystals—and put equal amounts into each solvent. Quite different results are obtained. This time the carbon(IV) chloride dissolves most iodine and water the least.

2.14 The density of a solid

The density of a solid—for example, an element, a compound or a mineral—is the ratio of mass to volume. The mass is readily determined by means of a balance. If the solid is insoluble in water, the volume can be found from the volume of water it displaces, regardless of the shape of the solid. Half fill a graduated cylinder with water. Note the reading. Immerse the solid in the water and note the reading again. The difference in the two readings is the volume of the solid. A few examples of substances and their densities (in $g\ cm^{-3}$) which can be of interest to the chemist are: sulphur: 2.0; quartz: 2.6; calcite: 2.7; copper: 8.9; lead: 11.4. Ores such as malachite, cassiterite and cerussite are not uniform in density as they contain variable quantities of quartz, feldspar and other minerals. (See also experiments 2.286, 2.287, 4.9.)

2.15 Density of a liquid

Toluene, carbon(IV) chloride and bromoform are interesting liquids to investigate. Feldspar and quartz will float on bromoform, density $2.9\ g\ cm^{-3}$. Weigh a small container with the liquid inside. Pour the liquid into a graduated cylinder to find the volume of the liquid. It will not matter if any of the liquid adheres to the side of the container. Use a balance to find the mass of the container and hence find the mass of liquid transferred to the measuring cylinder. Obtain the density by dividing the mass of the liquid by the volume. (See also experiments 2.286, 2.287.)

Energy to change solids to liquids and liquids to vapour

2.16 Investigating the heat energy when a liquid changes to a solid

Naphthalene is a suitable substance in warm climates. It is a liquid above 80.2°C. Benzene is more suitable for a cold climate because ice is needed to cool it below the melting point of 5.5°C.

Put some crushed naphthalene to a depth of 6 cm in a 100×16 mm test-tube. Heat gently until all the naphthalene has melted. Put a thermo-

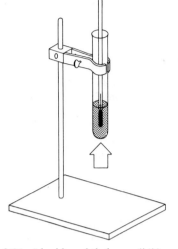

2.16 Liquid naphthalene solidifies

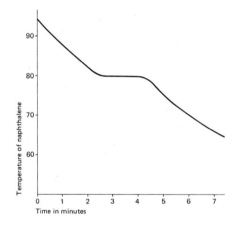

This experiment can be done by pupils using water and by the teacher using ethanol, trichloroethane or tetrachloroethane. The first essential is to have a very constant source of heat which is free from the effects of draughts.

A. A bunsen burner with a flame about 5 cm high is put below a flask or beaker containing a known mass of the liquid. A suitable volume is about 50 cm³. The set-up is shown in the diagram (note the draught shields around the apparatus).

Place a thermometer in the liquid and record the rise in temperature every 15 seconds. Plot the temperature against time on a graph. The best straight line that can be joined through the points obtained (omitting the later ones) has a slope from which the average temperature increase per minute can be calculated. Assume that all the heat goes into the liquid. The heat absorbed by the flask is small in comparison. The number of calories absorbed by the liquid per minute can be calculated by multiplying the mass of the liquid by its specific heat and by the temperature increase per minute. (See also experiments 2.135, 2.136.)

meter into the naphthalene. Stop heating when the temperature of the naphthalene is about 95° C. Have ready a clock or watch with a second hand. Stir the naphthalene gently with the thermometer whilst it is cooling. Record the temperature every 15 seconds. Continue taking readings for about 6 minutes.

Plot a graph of temperature against time as shown in the diagram. It can be seen that the temperature does not fall so fast at the melting point. Can pupils explain why the cooling is delayed for some seconds when the melting point is reached?

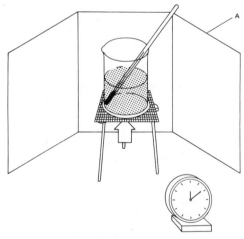

Finding the heat required to vaporize a liquid
A draught shields

B. Without altering the flame or shields in any way, allow the liquid to boil for a known time (ten minutes is suitable for water, but five minutes for the more volatile liquids). From the first part of the experiment, the amount of heat supplied to the liquid can readily be calculated for a specific number of minutes. Can pupils suggest what the heat has been used for? After the boiling, remove the bunsen, allow to cool and find the mass of the liquid which has been vaporized. The results so far enable the calculation of the amount of heat necessary to convert a certain mass of liquid into vapour. How many kilocalories are therefore necessary to convert 18 g water or 46 g ethanol, etc., into vapour? This will be the heat of vaporization of the liquid.

Applying melting point, boiling point, solubility and density to problems of separating substances from mixtures

2.18 *Separating tin from a mixture of tin and carbon*

Make a mixture of tin and carbon, using tin filings, or small cut pieces of tin, and crushed charcoal. Small pieces of tin solder could be used, although the composition of solder is only approximately 66 per cent tin, the rest being lead. Lead is an alternative. 'Tin cans' are useless, as these are iron with only a very thin surface of tin. Tin melts at 232° C, and carbon at 3,730° C. Heat the mixture in a crucible. Stir with a splint until the tin melts and forms a liquid below the charcoal. Pour the tin on to an asbestos mat or some other heat-proof surface, holding back the charcoal in the crucible by means of a wood splint. Alternatively the tin might be cast in a previously prepared plaster of paris mould.

2.19 *Separation by sublimation*

Separate iodine from a mixture containing a few crystals of iodine and sodium chloride. Heat the mixture in an evaporating dish with a funnel placed on top (this is shown in the diagram). The iodine sublimes on to the cool sides of the funnel.

2.19   Sublimation of iodine

2.20 *Separation by distillation*

This is a separation of water from ordinary ink. Pupils should understand that this is a process of evaporation of the more volatile water as vapour, followed by condensation of this vapour back to water in another vessel. It is important that a coloured solution like ink is used in order to give emphasis to the separation. Put about 5–10 cm³ of ink into a container for boiling, together with some boiling chips or granules. This container might be a small conical flask or a boiling tube. Fit a stopper with delivery tube to the container. The outer end of the delivery tube should go half-way down a test-tube as shown in the diagram A or into a U-tube as shown in Boverleaf. Cool the collecting tube. Heat the ink with a very small bunsen flame. Pupils can observe the

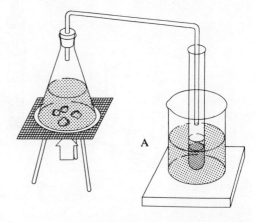

Condensing the vapour (see overleaf)

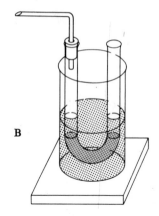

B

visible boundary between the hot and the cold parts travel up the boiling-tube or conical flask and along to the collecting tube. After a few minutes, drops of a colourless liquid appear in the collecting tube. The liquid may be identified as water by its action on anhydrous copper sulphate. Care should be taken to prevent ink from frothing up or splashing into the delivery tube.

2.21 *Separation of crude oil by fractional distillation*

Crude oil can readily be separated into three or four fractions each having interesting fuel or lubricating properties. A substitute for crude oil may be obtained by mixing suitable proportions of used car oil, petrol, paraffin, thin lubricating oil, diesel oil and a little vaseline. Set up the hard glass test-tube, the delivery tube and the five small ignition tubes as shown in the diagram.

Use a 0–360° C thermometer if one is available, but in this case a test-tube with a side arm as shown in the diagram at B is better than an ordinary test-tube as at A in the diagram. Put about 4 cm³ of oil into the test-tube. Either asbestos wool or boiling chips should be added to prevent bumping. Set up five small ignition tubes to collect the fractions. Heat the oil very gently. Collect about 10 drops of distillate in the first tube, then 10 drops of distillate in the second tube, etc. The boiling point of the remaining oil will become higher as distillation proceeds and more heat will then be required from the bunsen. Arrange the fractions in order of increasing distillation temperature. It should be possible to make the following observations.

1. The colour should change from colourless to yellow.
2. The viscosity should increase. 'Runniness' should decrease.
3. The high-temperature fractions should be more difficult to ignite than the low-temperature fractions.
4. The high temperature fractions should burn

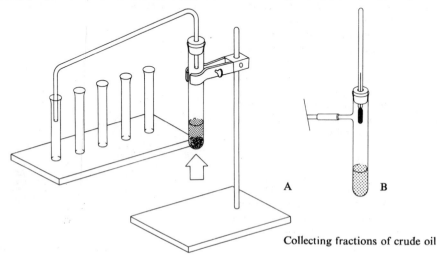

A   **B**

Collecting fractions of crude oil

with more soot in the flame than low-temperature fractions.

Bottle tops, with the cork removed from inside, make suitable containers in which to burn the fractions. After this experiment, which fractions would pupils select for use as petrol, paraffin, diesel oil and lubricating oil? What use could be made of the dark residue remaining in the test-tube?

If a thermometer (0–360° C) has been used, the fractions are collected over suitable ranges of temperature: (a) up to 80° C; (b) 80–120° C; (c) 120–180° C; (d) 180–220° C. With their knowledge that pure substances can be recognized by their boiling point, would pupils think that each fraction collected is a pure substance? Suggest that pupils search out information on oil refineries.

2.22 *Separating salt and sand*

Prepare a mixture of salt and sand. Put about 2 cm^3 of the mixture in a 100 × 16 mm test-tube. Add about 5 cm^3 of water and shake until all the salt has dissolved. Pour the contents of the tube into a filter paper held in a funnel and supported over an evaporating basin. Wash the test-tube with a little water and add this to the filter paper. The sand will remain on the filter paper and may be dried and collected. The salt can be recovered from the filtrate by warming the evaporation basin to drive off the water.

2.23 *Solvent extraction of oil from nuts*

Put about twelve ground nuts (or pieces of chopped coco-nut) into a mortar. Add 20 cm^3 of acetone or methylated spirits. Grind the nuts as finely as possible for a few minutes with the solvent. Then pour off the liquid into a test-tube. Filter the liquid into an evaporating basin. Leave the basin in a sunny place for about 5–10 minutes, or, if there is no sun, place it on top of a beaker of hot water for 15 minutes. The solvent will evaporate and leave the oil which has been extracted from the nuts.

2.24 *Chromatography as a separation technique*

A. Collect some leaves and grass. Dry them. Tear or cut the leaves and grass into small pieces. Put them in a mortar. Add about 5 cm^3 of acetone or alcohol. Grind the leaves and grass very thoroughly with the solvent until a deep-green solution is obtained. (The reason for not adding much solvent is that as concentrated a solution as possible is required.) Cut a strip of filter paper 1 cm wide and long enough to be suspended in a test-tube without touching the bottom. Using a fine dropping tube, put one drop of the concentrated solution on a point 1 cm from the bottom edge, as shown in the diagram at A. Wave it gently to dry it quickly. Then add another drop on the same spot. Dry this. Add more drops, each time drying the drop before another is added. The idea is to obtain a small concentrated spot of the coloured substances in the leaves and grass. Now put 1 cm^3 of the solvent in a test-tube. Hang up the strip of blotting paper in the test-tube with the bottom end just immersed in the solvent and with the spot A well above the solvent level. This is shown in the diagram. The solvent will be drawn up the filter paper by capillary attraction. The coloured substances will be taken up with the solvent to an extent depending on their distribution between the paper and the solvent. The chromatogram should show a top orange band of

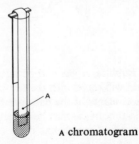

A chromatogram

xanthophyll and a lower green band of chlorophyll. If benzene or toluene is used as the solvent, a band of carotene should also be visible between the other two.

B. Coloured inks, particularly black ink, usually contain several dye colours. These colours can be separated by paper chromatography using the apparatus of the last experiment. Put one small drop of black ink at the point A. Hang the filter paper in an acetone/alcohol mixture made alkaline with a few drops of ammonium hydroxide. Try other solvents. A good separation of dye colours should be obtained.

2.25 *Finding out how much gas is dissolved in a sample of water*

Fill a round-bottomed flask right up to the top with tap water and insert a stopper with a delivery tube which is also completely filled with water. (An easy way to do this is to insert the stopper while holding the whole apparatus under water in a sink.) Set up the apparatus as shown in the diagram and heat the flask by means of a bunsen

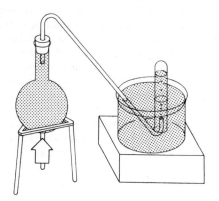

burner. Bubbles of gas will be released from the water and will travel into the test-tube. Continue until the contents of the flask are boiling. About half a test-tube of gas may be collected from a litre of water, all of which has been displaced from solution by heating.

2.26 *Separating two immiscible liquids of different density*

Suitable mixtures with their densities, d, in g cm^{-3}, are: (a) water and benzene ($d = 0.88$); (b) water and carbon(IV) chloride ($d = 1.59$); (c) water and mercury ($d = 13.6$).

Suitable vessels for separating one liquid from the other would be a burette or a piece of wide tubing fitted with a stopper, tube and rubber tube with clip as shown in the diagram. A single crystal of iodine added to mixtures (a) and (b) would show up the benzene and carbon(IV) chloride layers more clearly. Shake the mixture and run it into the separating vessel. Wait until a clear boundary appears between the two liquids and then run off the heavier layer into a beaker below.

2.27 *Separation of two solids by density differences*

In industry, diamonds (density about 3.3 g cm^{-3}) are separated from feldspar and quartz by floating the latter in a slurry of magnetic iron oxide made up to the appropriate density.

Bromoform is not very familiar to school laboratories but if it does happen to be there the following makes an interesting experiment. Beach sand often consists of quartz particles mixed together with heavier particles such as ilmenite or zircon. In bromoform, which has a density of 2.9 g cm^{-3}, the quartz particles will float and the heavy particles will sink. To 3 cm depth of bromoform in a test-tube, add a little sand. Is there any separation of quartz and heavier minerals? Float glass on bromoform. Can you find any rocky particles which sink in bromoform? The same bromoform can be used for many experiments. Do not throw it away.

The effect of heating substances

When a substance is heated several changes may be observed. It may melt, boil, change into a new substance, change on heating but re-form on cooling, change colour, change in volume, gain mass, lose mass, or it may remain the same mass.

2.28 *Substances which add on something from the air*

A. Clean a piece of copper foil about 3 cm square. Hold it in a pair of tongs and heat it. A black substance forms on the copper. Is this black substance something from the flame? Is it because something has been added to the copper from the air? Has the black stuff come from inside the copper? Can experiments be planned to answer these questions? Using a much larger surface area of copper, is it possible to find out if the copper changes in mass when heated?

B. An experiment with magnesium. Clean about 25 cm of magnesium ribbon. Cut this into pieces 1 cm long. Put them into a crucible with a lid. Weigh the crucible plus lid plus magnesium. Place the crucible on a pipe-clay triangle supported on a tripod. Heat very gently at first and then as strongly as possible. Hold the lid in a pair of tongs close to the crucible. The magnesium darkens just before it begins to melt. At the first sign of burning place the lid on the crucible and remove the bunsen. About every 4 seconds, raise the lid just a little to allow more air to enter. Try not to allow any white magnesium oxide smoke to escape. When the magnesium stops burning on raising the lid, remove the lid cautiously. Heat the crucible again strongly. Keep the lid ready in case the magnesium starts to burn again. Allow to cool. When cool, weigh the crucible plus lid plus contents. Is there any increase in mass of the magnesium? Where has the increase come from?

2.29 *Collecting and weighing gaseous products of burning*

Solid products resulting from burning are readily weighed. But what about the products which are gaseous?

To find if a candle takes something from the air, the gaseous products must be weighed. Candle wax, being a hydrocarbon, burns to water vapour and carbon dioxide. Soda lime granules will absorb both these gases. Assemble the apparatus as shown in the diagram.

First weigh the whole apparatus. Then turn on the filter pump to draw air over the candle, and light the candle. Allow the candle to burn for 5 minutes. Extinguish the candle and disconnect the water pump. When cool, weigh the whole apparatus again. Has it gained mass? Does the candle take oxygen from the air during burning? Is the increase in mass due to the water vapour being absorbed from the air being drawn through the apparatus? Pupils may like to repeat a control experiment without lighting the candle and drawing the air through with the filter pump at the

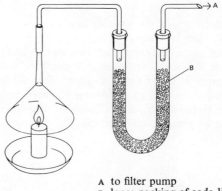

A to filter pump
B loose packing of soda lime

same rate and for the same length of time. In the experiment the candle by itself has lost mass, but the increase of mass of the U-tube due to the trapped gaseous products must be more than the loss in mass of the candle.

2.30 *Substances which lose mass when heated*

A. Weigh a test-tube containing potassium permanganate to a depth of 1 cm, and with a plug of cotton wool at the mouth to prevent loss of any solid during heating (see diagram overleaf). Heat the tube. Weigh it again. Is there a loss in mass? Where has it gone?

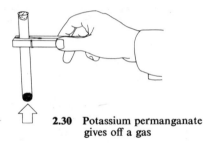

2.30 Potassium permanganate gives off a gas

B. Try heating copper carbonate in the same way. Does this lose mass? Where has the mass gone?

2.31 *Some substances neither gain nor lose mass when heated*

Heat dry zinc oxide in a test-tube in the same way as in the last experiment. Is something lost or something gained?

2.32 *Observing the effect of heat on copper sulphate crystals*

Crush some blue copper sulphate crystals and put them into a dry test-tube to a depth of 4 cm. Arrange as shown in the diagram. Heat the tube gently. What can pupils observe? Vapour collecting on the cooler parts? Change of colour from blue to white? Liquid collecting in the receiving

tube? Can the identity of the liquid be established by finding out the boiling point? When all the copper sulphate crystals have changed to white and the tube has cooled, hold the tube in your hand and pour the liquid back on to the white crystals. Is the blue colour restored? Is any heat given back again? One way of recording what has happened is to write :

blue copper sulphate crystals + heat $\rightleftharpoons$
white (anhydrous) copper sulphate + water.

This is a reversible change. Pupils might discuss whether the previous experiments on heating substances were reversible changes.

Preparing, collecting and testing some gases

2.33 *Hydrogen*

A. Place a few pieces of granulated zinc or zinc foil from the casing of an old dry cell in a boiling tube; add 2 drops of copper sulphate solution and assemble the apparatus as shown in the diagram. An alternative to the thistle funnel at A is a syringe as shown at B. Syringes may be obtained as used rejects from hospitals and clinics. Pour enough molar sulphuric acid (see Chapter One) down the thistle funnel on to the zinc to cover the

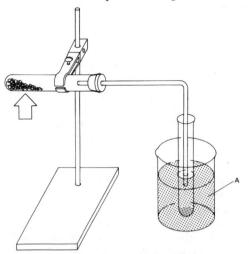

2.32 Collecting the product of heating copper sulphate crystals
A cold water in a beaker

2.33 Collecting hydrogen gas

bottom of the funnel tube. Alternatively, add some sulphuric acid from the syringe. In this case gas cannot escape through the syringe so that the tube of the syringe does not have to be covered by acid. Discard the first two or three test-tubes of hydrogen as they will contain displaced air. WARNING. There can be a dangerous explosion if any vessel bigger than a test-tube is used when igniting the gas particularly if it is mixed with air. Collect a few test-tubes of the gas and cork them. Test the third test-tube of gas by holding a lighted taper or splint over the mouth immediately the cork is taken out. Pure hydrogen burns with a quiet 'pop'. Hydrogen should never be dried with concentrated sulphuric acid.

B. Hydrogen burns in air to form water vapour. When hydrogen is ignited in a dry test-tube, can any vapour or mist be detected on the sides of the test-tube?

C. Investigate whether hydrogen is lighter than air by 'pouring' the gas into a test-tube held either above the first tube or below it. Use a lighted taper to find out where the hydrogen has gone. Blow some soap bubbles by holding the delivery tube of the apparatus in some detergent or soap solution. The hydrogen bubbles will rise into the air, again showing the low density value of hydrogen gas.

2.34 *A small hydrogen generator*
A simple apparatus for generating hydrogen is shown in the diagram. A is a boiling tube with holes drawn in the bottom. (The holes can be made by heating both the bottom of the test-tube and a glass rod to red heat in a bunsen flame. Fuse the glass rod on to the part where a hole is needed and then, by pulling the rod away, a shred of glass is pulled out from the boiling tube. Break this off and round off the edges in a hot flame. Make three or four holes like this in the boiling tube.) Put some granulated zinc metal into the boiling tube and cork it with a delivery tube and clip as shown. Immerse this in a jam-jar containing molar sulphuric acid plus a few drops of copper sulphate solution. When the clip is

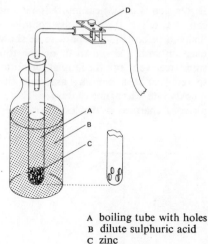

A boiling tube with holes
B dilute sulphuric acid
C zinc
D screw clip

opened acid will enter A and react with the zinc. When the clip is closed the pressure of hydrogen produced will force the acid from A through the holes and the reaction will cease. To prevent tiny pieces of zinc passing through the holes, cover the bottom with glass wool.

2.35 *Oxygen*
A. Oxygen can be prepared safely by decomposition of hydrogen peroxide solution. Hydrogen peroxide is often sold in chemist shops or drug stores. Put about 20 cm³ hydrogen peroxide into a bottle of about 100 cm³ capacity. Add two spatulas of manganese dioxide and fix a delivery tube to the bottle. Oxygen will bubble off and can be collected as shown in the diagram.

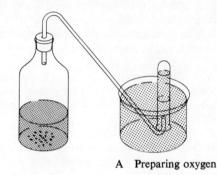

A   Preparing oxygen

B. Oxygen is colourless and has no smell. How can you tell if the test-tube contains oxygen? Shape a piece of nichrome wire as shown in the diagram with a shield to fit on the top. Fasten some steel wool into the loop at the bottom. Heat to red heat in a bunsen flame and then insert quickly into a test-tube of oxygen. Fasten a small piece of charcoal in the loop. Try to ignite the

2.36 Collecting hydrogen chloride

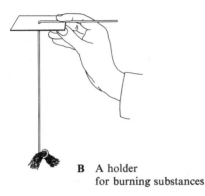

B A holder for burning substances

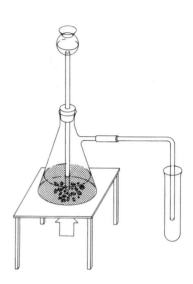

charcoal in the bunsen flame and then insert quickly into another test-tube of oxygen. Dip the loop into some sulphur powder. Ignite in a bunsen flame and then put into oxygen. What happens to these three substances? What happens to a glowing piece of string or a glowing splint of wood when they are put into a test-tube of oxygen?

2.36 Hydrogen chloride
Put some rock salt (sodium chloride) into a 100 cm³ filtering flask. (Coarse rock salt causes less frothing than the fine salt.) Carefully add concentrated sulphuric acid down the thistle funnel. Hydrogen chloride gas can be collected by upward displacement of air (see diagram).

A. Collect four test-tubes of the gas and cork them. Remove the cork from one of these test-tubes under water. How soluble is hydrogen chloride?

B. Hold a piece of cotton wool soaked in ammonium hydroxide at the mouth of a test-tube of

hydrogen chloride. The white cloud of ammonium chloride helps to identify hydrogen chloride.

C. Shake a test-tube of the gas with water to obtain a solution of hydrogen chloride. Test the solution with an acid/base indicator (see experiment 2.44). React a little magnesium with the solution. Can the hydrogen from this reaction be collected and tested?

2.37 Ammonia
A. Put a mixture of calcium hydroxide and ammonium chloride into a test-tube to a depth of 4 cm. Fill a U-tube with lumps of calcium oxide mixed with cotton wool. (The cotton wool is to prevent blocking of the tube.) Set up the apparatus as shown in the diagram. Gently heat the test-tube. The calcium oxide dries the ammonia gas. Test whether the receiver test-tube is full by holding a piece of red litmus paper at the opening. Collect test-tubes of ammonia and cork them. The method of collection illustrates that ammonia gas is lighter than air.

2.37A Preparing ammonia
       A calcium oxide lumps

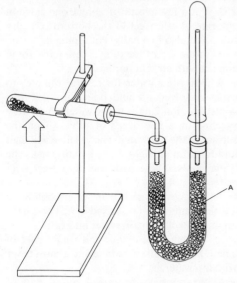

2.37B A fountain experiment

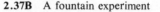

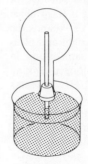

B. Fill a flask with ammonia. Fit a cork and tube into the flask as shown (ideally, the tube should have been drawn out into a jet). Warm the flask gently to expand the gas a little and then hold the flask upside down with the tube in the water. After a little time water will spray into the flask from the jet.

2.38 *Carbon dioxide*
Many reactions can be used to produce carbon dioxide gas. Marble chips or other carbonate rock

treated with dilute acid provides a good source. The gas is not too soluble to be collected by water displacement (as shown above for the preparation of hydrogen). Alternatively, carbon dioxide can be collected by displacing air from dry bottles as shown in the diagram (i). To test if the bottle is full, lower a lighted splint or taper into the top of the jar. If the flame is extinguished at the entrance as at (ii), then the jar is full. Put a cardboard cover over the top to prevent diffusion of the gas. Check the density of the carbon dioxide by 'pouring' the gas into another bottle either above or below the first bottle. Find where the gas is by testing with a lighted splint. *Note:* The presence of carbon dioxide can be confirmed by the fact that lime-water becomes 'milky' when the gas is passed through it.

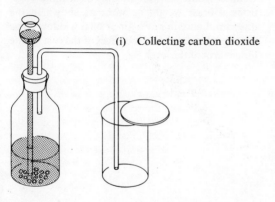

(i) Collecting carbon dioxide

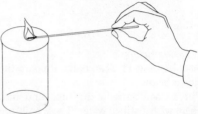

(ii) Finding out when the jar is full

2.39 *Cooking and carbon dioxide*
Pupils should understand that the purpose of baking powder or soda in cooking is to produce tiny bubbles of carbon dioxide. This expands the pastry, cake or dough, making it light and pleasant to eat. Yeast cells do the same thing in bread-making, though this takes longer. Baking powder (or sodium bicarbonate, $NHCO_3$) reacts with an acid such as lactic acid from sour milk to produce carbon dioxide. Commercial 'baking powders' often contain a solid acid which only reacts with the sodium bicarbonate when moist.

A. Put some baking powder into water. Is carbon dioxide gas given off? Is carbon dioxide given off if sodium bicarbonate is put into water? React baking powder in a test-tube with vinegar (acetic acid) or lemon juice. Is carbon dioxide produced? What kind of substance is lemon juice?

B. Make a sugar solution and half fill a jar with this solution. Add a spoonful of yeast. Leave this to stand for 2 or 3 days. Construct a 'bubbler' to go on the top of the jar as shown in the diagram. Is a gas given off by the yeast? Does carbon dioxide collect in the top part of the jar?

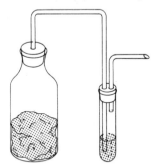

Yeast reacting with sugar solution

What is rusting?

2.40 *What causes rusting?*
Take 7 test-tubes and 11 clean nails. Prepare the tubes as shown below:
Tube 1: Put 2 clean nails in the tube and half cover them with distilled water. These nails are in contact with air and water and form the control experiment.

Tube 2: Put a few pieces of anhydrous calcium chloride or silica gel in the bottom of a dry tube, and also two nails. Put a plug of cotton wool in the top of the tube. These nails are in contact with air, but not moisture.

Tube 3: Boil some water for several minutes to expel dissolved air and pour into the tube whilst hot. Put 2 nails in the water. Put a little vaseline or a few drops of olive oil on the surface of the hot water. The vaseline will melt and form an air-tight layer, solidifying as the water cools. These nails are in contact with water but not air.

Tube 4: Half cover 2 nails with water containing a little common salt dissolved in it. These nails are in contact with air, water and salt.

Tube 5: Wrap a piece of zinc foil round part of a nail. Put the nail in the tube and almost cover with tap water.

Tube 6: Wrap a piece of tin foil round part of the nail. Put the nail in the tube and add tap water as you did for tube 5.

Tube 7: Wrap a piece of copper wire round a nail and put it in the tube exactly like tubes 5 and 6.

Stand these 7 tubes in a rack and leave for several days. What do the pupils conclude are the conditions for rusting? Which metal is better at preventing rusting, zinc, copper or tin?

2.41 *Does iron increase in mass during rusting?*
Counterbalance a piece of iron on a knife edge, using a brass weight or stone as shown in the diagram. Leave in moist air or on a window ledge for a few days and note the effect of the rust on the longer arm of the lever.

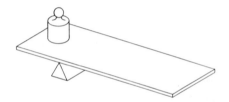

2.42 *Finding out what combines with iron during rusting*

Moisten the inside of a test-tube with water, sprinkle into it a spatula measure of iron filings and rotate it horizontally so that the filings spread and adhere to the walls. Alternatively, push a small plug of moistened iron wool to the bottom of the tube. Invert the test-tube in a beaker about one-third full of water. Use the beaker lip to support the tube as shown in the diagram. The water levels inside and outside the tube should be the same and the level should be marked on the tube. Leave the tube in this position for a few days. The iron will rust and the water level will rise up inside the tube, finally becoming steady.

Again add water to the beaker until the levels inside and outside the tube are the same and mark the new level. It will be seen that one-fifth of the air volume has been used up, suggesting that oxygen has been used up in the rusting of iron. The residual gas does not support combustion of a lighted splint. (See also experiments 2.318 and 4.58.)

Coloured extracts from flowers as indicators of acids and bases

2.43 *Extracting coloured substances from plants*

Select some brightly coloured flowers, such as the purple and red bougainvillea, or coloured leaves. Squeeze or grind one of the coloured flowers or leaves in a mortar with a mixture made of 2 cm³ acetone and 2 cm³ ethanol. By this means the colouring matter will be extracted into the solvent.

Filter and collect the filtrate. Repeat this experiment with one or two other flower colours. Keep these coloured solutions for use as 'indicators' in the next experiment.

2.44 *Using plant extracts to 'indicate' whether a substance is acidic or basic*

Put a spot of the coloured flower extract on to a filter paper and allow the paper to dry. Put one drop of lemon juice on to the spot. Is there a change of colour? Select other 'sour' fruit juices, bottled fruit juices and vinegar and carry out the same experiment. These are acidic substances. What is the colour change with dilute hydrochloric acid? The slightly different colours obtained would suggest that some substances are more acidic than others. Put some of the original filtrate on to another piece of filter paper. When dry, what are the colours given by sodium bicarbonate solution, washing soda, lime-water and a dilute solution of sodium hydroxide? These are alkaline, or basic, substances. Do they all give the same colour?

We have seen that plant extracts act as 'indicators'. They give an idea of how acidic or how basic a substance is. Add a few drops of sodium bicarbonate solution to 1 cm³ of flower extract indicator in a test-tube. Then a lemon juice and note the colour change. Try the same experiment with lime-water and indicator followed by dilute hydrochloric acid. What happens? Can the original colour be obtained by adding more lime-water? How many times can the indicator colour be changed before the test-tube is full? Litmus— an extract of lichens—is another plant indicator.

Chemists make up a Universal Indicator either as a solution or dried on filter paper. This Universal Indicator not only indicates whether a substance is acidic or basic but also how acidic it is. Pupils might investigate the effect of Universal Indicator on all the solutions mentioned. In order to avoid using the name of a colour to indicate acidity, a scale of numbers from 0 to 14 is used. This is called the pH scale and although it has a quantitative mathematical derivation, it may be used simply to express the degree of acidity and alkalinity as a number between 0 and 14. Acidity is the property of any solution which has a pH less than 7. Solutions with a pH greater than 7 are alkaline or basic. Solutions with a pH of 7 are neither acidic nor basic. They are neutral. Investigate the pH of water. Is water neutral? On the bottle or packet of the Universal Indicator there will be a chart giving the colour and the pH number associated with this colour. The colour changes on a simple Universal Indicator may be as shown:

Colour	pH number	Acid/Base
Red	1–3	very acidic
Orange	4–5	weak acid
Yellow	6	very weak acid
Green	7	neutral
Blue	8	very weak base
Indigo	9–10	weak base
Violet	11–14	very basic

Use 2 drops of Universal Indicator to 10 cm³ of solution to be tested.

Crystal growth

2.45 *Watching crystals grow*
Sodium thiosulphate crystals grow rapidly from a super-saturated aqueous solution. The formula for the crystals is $Na_2S_2O_3$ $10H_2O$. On heating, these crystals dissolve in some of their water of crystallization. Put sodium thiosulphate crystals into a test-tube to a depth of 3 to 4 cm. Add 1 or 2 drops of water. Heat gently until all the crystals have dissolved. They appear to 'melt'. Leave to cool. Crystals are unlikely to form unless a tiny seed crystal of sodium thiosulphate is dropped into the solution. When this is done, crystal growth commences and spreads rapidly through the whole solution. It is fascinating to watch the growth from one centre. Are pupils able to make any further comment if they hold the tube in their hand whilst crystallization occurs?

2.46 *Watching crystals of naphthalene grow from the melt*
Put a little naphthalene on a glass slide. Hold over a flame until the crystals melt. Put a cover slip over the liquid and allow to cool. Watch the crystals grow using a hand-lens. Sometimes crystals will grow from several points simultaneously and this will give rise to 'boundaries' where they meet. Pupils might attempt to draw the shape of the boundary between the forming crystals and the melt. It is striking to view the crystals through polaroid filters.

2.47 *Crystals with different shapes*
Experiment to find the correct concentrations of the following salts in aqueous solutions which will form good crystals when placed on a microscope slide. Too strong solutions will give clusters of crystals too quickly. Suitable solutions in test-tubes can be kept warm in a beaker of hot water during the investigation. The following are examples of different shapes in crystals:
Cubic crystals: sodium chloride and potassium chloride.

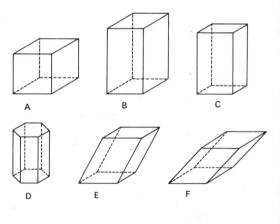

A regular (cube)
B tetragonal
C orthorhombic
D hexagonal
E monoclinic
F triclinic

Tetragonal crystals: nickel sulphate, potassium nitrate, zinc sulphate.

Monoclinic crystals: potassium chlorate, sodium sulphate.

Triclinic crystals: copper sulphate.

Octahedral crystals are formed when sodium chloride crystallizes from alkaline, urea or ammonium hydroxide solutions. Funnel-shaped crystals form from a mixture of sodium chloride and alum solution. Observe these crystals by putting 2 or 3 drops of the warm concentrated solution on to a microscope slide and viewing with a hand-lens or microscope.

A cloth cover held in place by wire
B seed crystal

2.48 Investigating the two different shapes of sulphur crystals

Both rhombic and monoclinic crystal forms of sulphur may be grown from a solution in xylene. Xylene is inflammable but it can be heated quite safely in a pyrex boiling tube over a small flame. Xylene boils at about 140° C, depending on the isomer proportion. This is above the melting point and transition point of sulphur. Toluene is an alternative solvent but as it is much more volatile and again inflammable, great care must be taken to prevent the vapour igniting. The boiling point of toluene is 111° C, which is again above the transition temperature of monoclinic and rhombic sulphur, which is 95° C.

Dissolve crushed sulphur in hot xylene, leaving an excess of sulphur at the bottom. On leaving to cool, the solution may become cloudy, but soon after the sulphur at the bottom solidifies and long needle-like crystals of monoclinic sulphur grow upwards through the solution. Cool the solution; it should still be a pale yellow colour due to the saturated solution of rhombic sulphur remaining. Pour one or two drops of the clear solution on to a microscope slide. Crystals of rhombic sulphur will grow and their difference in shape from monoclinic crystals can be observed using a hand-lens.

2.49 Growing large crystals

(a) The starting point for growing large crystals is a 'seed' crystal which should be from 0.5 to 0.8 cm long. These can be prepared by slow evaporation of about 30 cm³ of the saturated solution in a glass dish. Dry and tie a piece of cotton around selected seed crystals. Hands should be washed free of impurities before doing this because impurities easily affect the size and shape of the crystal. Hang the seed crystal about 5 cm from the base of the container by means of a wire bent as shown in the diagram. Fill the jar with a solution of the salt a little under saturation strength before you put the seed crystal in position.

(b) Another way of holding a seed crystal, and also one which provides a mounting for the crystal when it is grown, is to support it at the end of a glass tube. Take a length of glass tubing, about 3-mm bore, and heat in a flame until the end softens sufficiently to squeeze with pliers or tweezers so as to make a smaller hole. When cool, drop in seed crystals until one catches in this hole. Keep this one in place by dropping other crystals on top. Support the tube so that the seed crystal at the end is immersed in the growing solution. The seed crystal should then grow.

If the crystal grows disproportionately, or little crystals grow on the surface, then screw the cover on to the jar for a while. This will cause the little crystals to dissolve. If the crystal is not suspended it is best to turn the crystal every so often so that growth on all faces is equal. If a crystal is grown in a crystallizing dish the liquid may 'creep' up the sides. This can be prevented by rubbing vaseline round the upper inside rim. Evaporation may be increased by sitting the crystal growing jar on a tin with a 5-watt bulb mounted inside it. An air flow over the solution surface given by a fan will also hasten crystal growth.

Crystals grown from aqueous solution can be preserved by immersion in carbon(IV) chloride, benzene, or some similar liquid; or they can be preserved by painting with a clear varnish.

2.50 *Display clusters of crystals*
A crystal blossom. Soak pieces of charcoal, brick or unglazed porcelain in a saturated solution of sodium chloride. Keep the pieces covered by adding more saturated sodium chloride solution over a period of two weeks. At this point mix some prussian-blue dye or ink with the sodium chloride and add this to the pieces of charcoal. Then leave to evaporate to dryness. Blossoms of crystals will form. A variety of colours may be produced by adding different dye compounds.

A crystal crown. Cut a crown from a piece of tin taken from a fruit can. Fasten it with a piece of wire as shown in the diagram. Wrap the crown with strips of cotton cloth. Dip the whole crown into a solution of potassium dichromate and then leave to dry. Seed crystals will form on the cloth. Prepare a saturated solution of potassium dichromate at 80° C and immerse the crown for a day or so in this saturated solution. Red crystals should

form and make a beautiful display on the crown. If the crown is small, only a small amount of potassium dichromate will be required.

2.51 *Splitting crystals*
If crystals of calcite or sodium chloride are available, they can be split apart in the following way. Obtain a razor blade of the kind shown. A biologist's scalpel is also very suitable. Place the blade

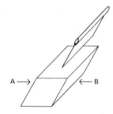

on the crystal with the face of the blade parallel to the planes of the two opposite sides A and B. Give a sharp tap with a small hammer on the top of the blade. Try to split the crystal with the least possible force. The crystal should split down the plane of symmetry. If the blade is not directed correctly the crystal will crumble instead of splitting into two parts.

The mineral galena (lead sulphide) occurs as cubic crystals which are easily split along the three cleavage planes at right angles to each other. The micas have one beautiful cleavage, and can be split into very thin flexible sheets.

Matter as particles: movement, number and size

2.52 *Brownian movement*
A. Colloidal graphite (available commercially as Aquadag) or toothpaste (which contains particles of magnesium oxide) may be used to observe the movement of very small, but visible particles suspended in liquids. Put a tiny drop of Aquadag on a microscope slide and stir distilled water into it until it is almost colourless. Alternatively, do the same with toothpaste. (Only the slightest trace of toothpaste is needed in the water.) Place a cover slip over the slide and put the slide on to the microscope stage. Illuminate from the side and observe with the lens of greatest magnifying power. It may take a little time to see the very slight jogging about of the particles. Select one tiny particle and keep your attention fixed on it. Although the particle at first appears to stay in one place, it is being continuously displaced in all directions. The reason for this is that it is being hit on all sides by the very much smaller and invisible molecules of water.

A model, which although not exact, may help pupils to understand what they are observing can be made by using a tray in which are placed a large number of small, light beads with one large marble in the middle. The small beads represent molecules of water and the large marble a particle of suspended graphite or toothpaste. Our model is not to scale since, in nature, the smallest visible particle (even under the microscope) contains some 10^{10} or 10^{11} atoms. When the tray is shaken, the small beads hit the marble many times from all directions. In this case, the forces cancel out within a small interval of time and the result is that the marble jogs about with very small movements but returns to the same place.

B. Fill a beaker with tap water and focus sunlight into the beaker by using a hand-lens. The motions of the suspended particles of solid matter may be observed where the light beam is brought to focus.

2.53 *A heavier-than-air gas diffuses upwards*
A. Fill a jar with carbon dioxide and invert it over a similar jar full of air. After a few moments separate the jars, pour a little lime-water in the lower one and shake it. The lime-water will turn milky indicating that the carbon dioxide has fallen into the lower jar because it is the heavier gas.

Now repeat the experiment but this time put the carbon dioxide in the lower jar and invert a jar of air on top of it. This is shown in the diagram. If the jars are left for about 5 minutes some carbon dioxide will be carried into the upper jar by diffusion; in the same way some air will be carried into the lower jar. The lime-water test will show the presence of carbon dioxide in the upper jar.

2.54 *Comparing the diffusion rate of ammonia and hydrogen chloride gases*
The apparatus is shown in the diagram. The long glass tube should be horizontal. Corks should fit at both ends. Using a pair of tongs or tweezers, dip a piece of cotton wool into concentrated hydrochloric acid and another piece into concentrated ammonium hydroxide. Drain off excess liquid. As nearly as possible at the same time, put the ammonia cotton wool at one end of the tube and the acid cotton wool at the other. Close the ends of the tube with corks. After a while, look carefully for a white ring which will form where the ammonia gas and the hydrogen chloride gas meet after diffusing through the air towards each other. Ammonia is the less dense gas and the white ring of ammonium chloride should form nearer to the hydrogen chloride end than from the ammonia end of the tube.

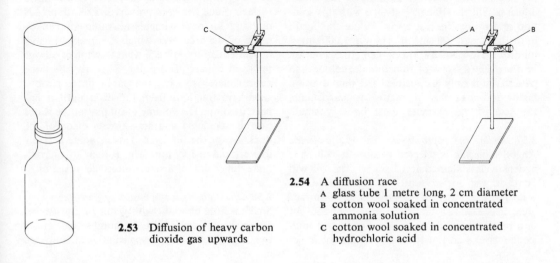

2.53   Diffusion of heavy carbon dioxide gas upwards

2.54   A diffusion race
   A   glass tube 1 metre long, 2 cm diameter
   B   cotton wool soaked in concentrated ammonia solution
   c   cotton wool soaked in concentrated hydrochloric acid

2.55 *Diffusion of liquids.*

A. Place a crystal of potassium dichromate or
ammonium dichromate at the bottom of a beaker
of water. One way of doing this is to put a glass
tube into the beaker of water so that it touches the
bottom, and then to drop the crystal down the
tube. Close the top of the tube with your finger
and remove the tube gently, leaving the crystal in
the beaker. The colour of the dissolving crystal
will spread throughout the water in quite a short
time.

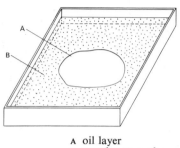

A   oil layer
B   powder on surface of water

B. Fill a very small open bottle with a strong
solution of potassium permanganate. Place this
in a larger jar. Fill the larger jar very carefully
by pouring water down the side until the water
level is above the top of the small bottle. Leave
this for a few days. The potassium permanganate
solution should have diffused evenly throughout
the water.

2.56 *Investigating particulate matter*

Put one crystal of potassium permanganate
in a test-tube. Add 1 cm³ of water. Dissolve the
crystal completely by shaking vigorously with
your thumb over the end of the test-tube. Then
add water to a total volume of 10 cm³. This is
a '10 times' dilution. Pour this 10 cm³ of purple
solution into a 100 cm³ beaker and then fill up
the beaker with water. This is now '100 times'
dilution. Fill the 10 cm³ test-tube with this solution and throw the rest away. Dilute this again
in the beaker to 100 cm³. It is now a '1,000 times'
dilution. How many times can the solution
be diluted by a factor of 10 before the colour is so
pale that it is only just visible? The final dilution
factor indicates that if matter is particulate,
the size of the particles must be very small.

2.57 *Finding the approximate size of a molecule*

An oil molecule is selected because it will have
a density less than that of water. The oil will
float on the surface and not dissolve in the
water. If the water has a large enough surface
area, it is assumed that thin oil will spread out
in a layer one molecule thick called a monomolecular layer and not form little 'hills' of mole-

cules. If the volume of oil is known and also
the surface area that it forms, then the thickness
of a monomolecular layer can be found by
dividing the volume by the area. A tray or container for the water is needed, which should not be
less than about 30 cm square so as not to restrict
the oil film. Sprinkle the surface of the water
with a very fine light powder such as talc powder.
When the oil is put on the water it will push
the powder away and the area covered by the
oil will be seen clearly (see diagram). To find
the volume of oil, pour thin oil into a burette
(a thin petroleum distillate is best). Find the
volume of fifty drops by running oil from the
burette drop by drop and counting the drops.
Allow one more drop to fall on a piece of plastic.
The point of a glass rod is allowed to touch the
oil drop and then it is touched on the water
surface which has been prepared. The oil spreads
out and an approximate measurement can be
made of the area over which it spreads. Finally
it is necessary to estimate what fraction of oil was
removed by the glass point. This may be done
approximately by using the glass point to remove
successive fractions from the drop until it has
been used up. The volume of oil put on the water
can be calculated and an estimate made of the
thickness of the oil layer. This estimate is likely
to work out at 10^{-6} mm. This is then an approximate dimension of a single molecule of the oil.

2.58 *Investigating a suspension of particles*

Shake a little clay soil with water in a test-tube.
Leave this to settle. Note the humus layer at the
top, the cloudy clay suspension next, and the

tiny particles of rock and mineral at the bottom. Filter the liquid. Pupils will observe that the filtrate is still cloudy; this is because the clay particles have passed through the filter paper. Do pupils understand why the suspension particles do not settle, even after a few days? The size of colloidal particles is roughly between 1 mμ and 100 mμ (1 mμ is 1 millimicron, which is 10^{-6} mm).

Divide the filtrate into parts in test-tubes. Keep one as a control. To the other add a few drops of barium chloride solution, or some aluminium salt solution. Note what happens in half an hour and in one hour. The same effect occurs when a clay suspension in a river meets the salts contained in sea-water. In many hot countries salt is crystallized from pans built on the clay beds near the mouths of rivers.

Electrical conductivity of substances

2.59 Solids which conduct electricity

To test for conductivity an apparatus such as that shown in the diagram may be used by pupils. The source of the d.c. supply can be dry cells in series giving 6 volts. The bulb, which should be low power, indicates when the current is flowing. The electrodes may be carbon or steel, perhaps mounted in a wooden support, cork or rubber bung so as to keep the electrodes a constant distance apart.

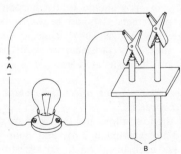

A   6 volt d.c. supply
B   electrodes

A. Test the conductivity of solids by making a good contact between the surface of the solid and the two electrodes (the surface of the solid must first be cleaned). Pupils might list all the metals they can find. All metals conduct electricity. Carbon conducts electricity. Do pupils find that non-metallic solids like plastics, naphthalene, wax, sugar, sodium chloride and sulphur do not conduct electricity?

B. Glass can be a conductor. Heat a glass rod until it becomes very hot and begins to soften. Test the hot, soft part with the conductivity apparatus. When molten, glass is a good conductor of electricity. (See also experiment 2.155.)

2.60 What liquids conduct electricity?

A. First test liquids obtained by melting substances. Melt the following substances, but heat very gently and cautiously because otherwise they may ignite and burn: sulphur, wax, naphthalene, polyethylene material, tin, lead and, if available, a low-melting point salt such as lead bromide (m.p. 488° C) or potassium iodide (m.p. 682° C). Test the conductivity of the melt by dipping in the electrodes and waiting a few moments for the electrodes to reach the same temperature. This ensures that the electrodes are in contact with the liquid and not the solidified melt. Scrape and clean the electrodes between each test.

B. Test ethanol (or methylated spirits), acetone, carbon(IV) chloride, vinegar, sugar solution, copper(II) sulphate solution, sodium chloride solution, and other substances dissolved in water. Clean and dry the electrodes between each test.

C. Test pure distilled water for conductivity. Put the electrodes into a beaker of distilled water. Pupils find that the bulb does not light up and therefore pure water does not conduct. Very gradually stir small crystals of common salt into the water. What happens to the bulb as the salt dissolves?

Are pupils now able to classify substances into

the following groups: (a) those which conduct electricity in the solid state and those which do not; (b) those which conduct in the liquid state and those which do not; (c) those which conduct when dissolved in water and those which do not?

Construction materials

2.61 *Preparing lead tin alloys*
The lead and the tin should be moderately pure. Tin melts at 232° C and lead melts at 327° C. Weigh out pieces of lead and tin to make four alloys: 20 % tin, 40 % tin, 60 % tin and 80 % tin. These are given as the percentage of tin by weight in the alloy. For each alloy, put the correct weights of lead and tin in a crucible or in a pyrex test-tube. Cover the metals with some carbon (crushed charcoal) to prevent oxidation of the metals. Heat the metals until they are molten. Stir the melt with a wood splint to help the metals dissolve. The molten metal is then cast into a mould.

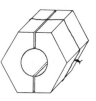

A suitable casting mould can be made by drilling out the thread of a nut to leave a smooth hole of about 0.6 cm diameter and then cutting the nut into two halves with a hack-saw (see diagram). Bind the two halves together with wire for casting and place them on a piece of asbestos. Pour the molten metal very carefully into the mould until full. Hold back the carbon with a wood splint whilst pouring. When cool, knock the two halves of the nut away from the cast alloy.

2.62 *Finding out whether lead tin alloys are harder than the pure metals*
Cast the four alloys and the two pure metals described in the preceding experiment, and label them with their percentage composition.

Obtain a metal punch with a good pointed end and a metal or plastic tube into which the punch fits loosely. The tube should be about 1 m long. The purpose of the tube is to guide the punch as it falls freely on to the surface of the metal alloy (see diagram). The sharp point of the punch will make a small hole in the surface of the alloy. With a softer alloy the hole will be larger. By measuring the diameter of the hole made by the punch, a comparison of

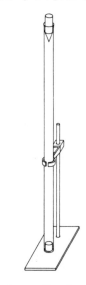

the relative hardness of each alloy can be made. The 60 % tin alloy should be the hardest alloy and the pure metals should prove to be less hard than these alloys. The diameter of the hole can be measured satisfactorily with vernier callipers and using a hand-lens.

2.63 *How does alloying affect the melting point of the metal?*
(a) Melt the four alloys and the two pure metals described in experiment 2.61, and pour a few globules of each metal on to an asbestos mat. Label each group of globules with the composition of the metal.

(b) Next prepare a piece of iron metal about

12 cm square and 0.2 to 0.4 cm thick. Mark the centre of the metal and draw a hexagon on the metal as shown in the diagram. Drill a small depression at each corner of the hexagon, each depression being the same size. Each depression will be the same distance from the centre. Drill small holes at the four corners, thread wire through them and suspend the metal plate horizontally by means of a stand. Use chalk to mark the six depressions on the plate with the following: 1. pure lead; 2. 20 % tin; 3. 40 % tin; 4. 60 % tin; 5. 80 % tin; and 6. pure tin.

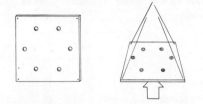

(c) Select pellets from globules made in (a) and corresponding in composition with 1 to 6 above. Put each pellet into the correct depression (see diagram). The metal plate should be heated exactly in the middle so the heat going to each pellet is the same. By prodding the pellets gently with a wood splint it can readily be seen whether they have melted. When all the pellets are molten, the splint may be used to remove molten metal from any of the bigger pellets so that they are all the same size. While the pellets are molten, remove the bunsen flame. Allow to cool. The first to crystallize will be the pure lead, having the highest melting point. Obtain a clock with a seconds hand. Start the clock (or note the time) when the lead solidifies. Note the time when each alloy and the pure tin solidify. Results may be tabulated:

Metal or alloy	Time, in seconds, from when lead solidifies to when alloy solidifies
Pure lead solidifies	Zero seconds
20 % tin solidifies	. . . seconds after lead
40 % tin solidifies	. . . seconds after lead
etc.	etc.

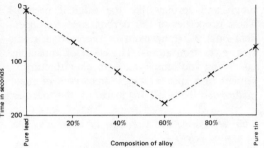

(d) Plot the results in graph form (see figure). It can be assumed that the alloy which takes the longest time to solidify after lead solidifies has the lowest of the melting points. The conclusions to be drawn from the last two experiments are that alloys containing about 60 % tin have the lowest melting point and they are also the hardest alloys.

2.64 *The effect of heat treatment on steel needles* Obtain some sewing needles about 4 to 5 cm long. These needles are alloys of iron and carbon, but the proportion of carbon is very small. Try bending a needle. It is tough and springy.

A. *Annealing.* Heat a needle to bright red heat. Hold it vertically in the flame and then very slowly raise it out of the flame taking about one minute. When it is cool, try bending it. It should be soft and easily bent round a pencil.

B. *Quenching.* Heat a needle to bright red heat and, whilst it is still hot, plunge it completely into cold water. Try to bend it now. It should be brittle and easily broken into small pieces

C. *Tempering.* Neither the soft needle nor the brittle needle are very useful. But the tough springy form can be restored. Heat and quench a needle as before to obtain the hard, brittle form. Carefully clean and shine the surface with emery cloth. The needle must now be heated very gently until a deep blue oxide film appears on the surface. This colour is an indication of the temperature at which the needle is tempered. When cool, try bending it. Is it tough and springy like the original needles? These properties of this carbon steel are dependent on the arrangement of the carbon atoms among the iron atoms. The effect of annealing, quenching and tempering is to alter this arrangement in a specific way. Some types of razor blades can be used in place of the needles.

2.65 *Comparing the strengths of mud, clay and sand bricks*

Find a source of clay soil or mud. If it is dry it must be mixed with water. To do this, put about 350 cm³ of water in a suitable container such as a plastic bowl. The best method is to crush the dry clay to a powder and then work it into the water until a thick smooth paste is obtained. Work it through your fingers until there are no lumps. It is likely to be of the correct consistency when it is thick and pliable, and sticks more to itself than to your fingers. Spread the clay or mud on to a flat surface very evenly so as to make a slab of 1.5 cm thickness. Then with a clean wet knife, cut 3 or 4 bricks, each the same size, 10 cm by 5 cm. Dry one under the sun for two or three days and bake another by a fire. Try making a sand brick of the same size. Also obtain a brick sold by a building contractor. Investigations with these bricks might include: (a) looking for cracks; (b) seeing if the surface comes away by rubbing with a dry finger; (c) seeing if the surface comes away by rubbing with a wet finger; (d) testing the strength of the small 5 × 10 × 1.5 cm bricks by supporting the two ends of the brick on the edges of two tables and loading the middle of the brick. This can be done by weights or, for strong bricks, by suspending a bucket into which sand can be poured

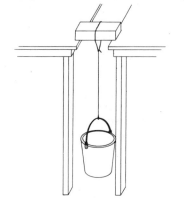

until the brick breaks. The weights and the bucket should be suspended near the floor as shown in the diagram so that they have only a little distance to fall.

2.66 *Using cement for making bricks*

Make 5 boxes out of stiff paper or cardboard 1.5 cm deep, 5 cm wide and 10 cm long. Use Sellotape or clips to fasten the edges (see diagram). A cement brick, the same size as the clay bricks, can be cast in these boxes. Smear a little oil or grease around the inside surfaces of the boxes. Obtain some fresh Portland cement from a builder.

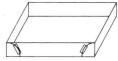

A. *A cement/water brick.* Mix the cement with water to a thick paste and fill the box with it, smoothing off the top surface level with the paper. It should 'set' in a few minutes but it will take a few days to 'harden'. 'Setting' is to change from a fluid to a firm rigid material but a mark can still be scratched on the surface with a nail. To 'harden' is to become rockhard.

B. *A cement/sand/water brick.* Mix 1 part of cement powder with 3 parts of clean sand. Work into a thick paste with water. Pour into the paper box, smooth off the surface and leave to set and harden.

C. *A cement/sand/gravel/water brick.* Make a brick as before using 1 part cement powder, 1 part of sand, 3 parts clean gravel and water. Cast the brick and leave to set and harden. This is a concrete brick.

D. *A cement/lime/sand/water brick.* A builder buys quicklime and mixes this with water to make calcium hydroxide on the building site just before he uses it. Mix 1 part of cement, 5 parts of builder's lime (calcium hydroxide) and 2 parts of sand and make into a paste with water. Cast a brick as before and leave to harden.

Suggest that pupils find out the kinds of

bricks used in their environment. They could make out a table of the properties and uses of the bricks. In a village the walls of houses may be made of mud or clay by mixing suitable soil with water. This may be plastered on to interwoven sticks built up as a wall; or bricks may be made with mud to build a wall. Such walls suffer damage during heavy rains. Sometimes the wall is covered with a cement and sand plaster inside and outside. The main structure of a modern building in a town is generally built of reinforced concrete. This is covered with a cement and sand plaster which has quite a rough surface. The smooth surface for inside buildings is obtained by using lime plaster.

2.67 *An experiment with plaster of Paris*
This is hydrated calcium sulphate. When mixed into a paste with water it sets quickly and expands. It is used as a fine casting material. Put 4 cm³ of water into a beaker. Add the powdered plaster of Paris slowly by means of a spatula. Continue adding the plaster until it just appears above the surface of the water. The plaster absorbs the water and you should finish with a very thin layer of water (about 1 mm) above the plaster. Stir the mixture well. When it begins to thicken, pour it into the paper box and smooth off the top as with the bricks mentioned in the preceding experiment. Leave to set for one day. Investigate the surface and strength of these bricks in a similar way to the investigations with the mud and clay bricks. Plaster of Paris is not often used as a construction material, but calcium sulphate as gypsum $(CaSO_4 2H_2O)$ is, in fact, used in the preparation of Portland cement.

Electrolysis of melts and aqueous solutions

2.68 *Electrolysis of a melt*
There are very few suitable low melting point salts. Lead bromide has a low melting point and, if available, makes an interesting electrolysis experiment. Potassium bromide may have too high a melting point (682° C) to melt easily. The apparatus is shown in the diagram.

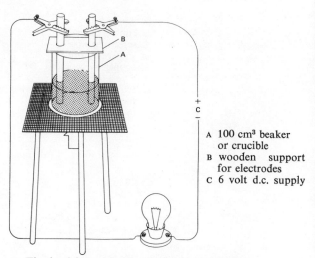

A 100 cm³ beaker or crucible
B wooden support for electrodes
C 6 volt d.c. supply

The lead bromide is melted in a small 50 or 100 cm³ hard-glass beaker, or in a crucible. The carbon electrodes are supported by a strip of wood with two holes bored 2 cm apart for the electrodes. Connect crocodile clips to the rods and complete the circuit with a torch bulb, to indicate when a current is flowing, and a 6- to 12-volt torch battery or some cells wired in series. The electrodes can be labelled positive and negative.

The only ions present in this melt are the bromide and lead ions. Bromine is readily seen coming off at the positive electrode which is the anode. The fact that bromine appears only at the positive electrode helps in the understanding of the existence of a negative bromide ion. Lead has both a lower melting point and a greater density than lead bromide and therefore appears as a melt at the bottom of the beaker. The small globule of lead which accumulates at the negative electrode (the cathode) can be seen after about 10 or 15 minutes of electrolysis. Decant off the molten lead bromide carefully into another crucible. The electric current has split up crystalline lead bromide into bromine gas and lead metal.

2.69 *Electrolysis of an aqueous salt solution*
Pupils should understand that in aqueous solutions there are usually four ions present, two

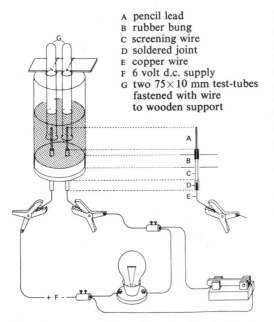

A pencil lead
B rubber bung
C screening wire
D soldered joint
E copper wire
F 6 volt d.c. supply
G two 75×10 mm test-tubes
   fastened with wire
   to wooden support

Drill two holes in the rubber bung with a 1 mm drill ($\frac{1}{32}$ inch). Insert the copper wire into the hole from above and pull it right through the bung until the screening wire is also pulled a little way into the hole. Into the core of this screening wire insert the pencil lead securely. Then pull the screening wire with the lead further into the bung so that the lead electrode is firmly held in the bung. The excess copper wire is cut off. This procedure is repeated with the other electrode.

Pupils should preferably make up the solutions for electrolysis to emphasize the fact that both a salt and water are present. The solution is placed in the glass cylinder. The two small tubes are then filled with the solution, and carefully inverted over the electrodes. The electrodes are connected to a safe d.c. supply with a small bulb in series. Increase the voltage until the bulb lights, showing that a current is flowing. When this happens, cut out the bulb from the circuit by closing the switch, as shown; this will allow a larger current to flow. The tubes collect any gas given off and the properties of the gas should be tested. Using carbon electrodes, the following results will be found.

A. *Electrolysis of water.* Pure water does not conduct electricity. For this reason 2 or 3 cm³ of dilute sulphuric acid or dilute sodium sulphate solution is added to the water in the electrolysis cell. Connect the cell to the d.c. supply and watch for bubbles of gas at both electrodes. If none appear, add a little more acid or sodium sulphate solution. After 5 to 10 minutes there should be enough hydrogen and oxygen gas for testing. Are pupils able to predict at which electrode each gas will appear?

B. *Electrolysis of solutions of ionic salts.* Most ionic salts can be used satisfactorily in electrolysis. Concentrations of 1M or less are suitable. (see Chapter One). Potassium iodide gives iodine at the anode and hydrogen gas at the cathode. Zinc sulphate gives a spongy mass of zinc at the cathode, oxygen gas at the anode. Lead acetate deposits lead on the cathode and oxygen gas is produced at the anode. (If the lead acetate solution is cloudy, add a few drops of acetic acid.)

from the water and two from the dissolved salt. The products will be gaseous, or metals which are deposited on the negative electrode.

The apparatus shown in the diagram is readily made. It is composed of an open cylinder of glass approximately 8 cm high and 2.5 cm in diameter. A small bottle of similar size with the bottom cut off would do just as well. The cylinder has a two-hole rubber bung carrying two carbon electrodes with connecting leads to a battery, or d.c. supply of 4 to 6 volts. If cork is used, this must be made leak-proof by covering the whole of the bottom surface round the electrodes and the glass edge with Faraday's wax or a similar soft wax. The electrodes may be carbon rods from a dry cell or pencil leads. The alloy supports for the coiled filament in electric light bulbs have also been found suitable for electrodes. The electrodes should project about 2 cm into the cylinder and also 2 cm below for attaching the leads to the battery.

Pencil leads are brittle, and if they are used it is better to fix the electrodes in the following way. Solder a piece of stout copper wire to a 4 cm length of braided copper screening wire.

Sodium chloride gives hydrogen gas at the cathode and chlorine gas at the anode. Copper sulphate deposits copper at the cathode and oxygen gas is produced at the anode.

Chemical reactions

2.70 A reaction between two elements
There have been examples in preceding experiments such as the reaction between oxygen and carbon, oxygen and sulphur, oxygen and copper. This experiment is a reaction between iron and sulphur. Mix together a spatula measure of sulphur with a similar amount of iron filings. Heat a small portion of the mixture on asbestos paper or a metal bottle top with the cork removed. Note that the reaction needs heat to start, but once started it continues without further heating. Test the product which is iron(II) sulphide. How different is it from the original elements? Powdered copper or zinc will react with sulphur using the same procedure as that outlined for iron. *Caution.* Use small amounts, because the reactions are usually vigorous.

2.71 Reactions between ions in aqueous solutions
A reaction between ions is readily shown by precipitation of an insoluble salt. A data book will give solubility data on salts. Examples of insoluble coloured salts are silver chromate, lead iodide and copper carbonate. To obtain these, dilute-aqueous solutions of the soluble salts shown in columns 1 and 2 below should be mixed in test-tubes. In each reaction, two ions, one from each salt, will form an insoluble precipitate as shown in column 3.

Pupils should be able to use the solubility information in a data book to obtain precipitates of silver iodide, barium sulphate, iron(III) hydroxide, and others.

2.72 Displacing copper from an aqueous solution of copper ions
A metal higher in the activity order is needed to displace copper metal from a solution of copper ions. Place approximately 10 cm³ of molar copper sulphate solution in a small beaker. Clean some magnesium ribbon and cut it into pieces 0.5 cm long. Add these pieces to the copper sulphate solution one at a time. The reaction can be vigorous. Copper metal deposits and the blue colour gradually disappears as the copper ion is displaced by the magnesium. Is any heat given out by the reaction? When the solution is colourless, decant the solution from the red copper powder at the bottom of the beaker. Collect the copper and dry it. How could the pupils show that this was copper and not magnesium? The usual way of writing the equation is:

$$Mg(s) + Cu^{2+}(aq) \rightarrow Mg^{2+}(aq) + Cu(s)$$

Pupils should repeat the experiment attempting to displace copper metal by using zinc and iron metal. The powder from zinc and iron is probably the best to use. Pupils should be able to discuss the comparative activity of the metals in this experiment.

	1	*2*	*3*
(a)	silver nitrate (aq)	+ potassium chromate (aq)	→ silver chromate (s)
(b)	lead nitrate (aq)	+ potassium iodide (aq)	→ lead iodide (s)
(c)	copper sulphate (aq)	+ sodium carbonate (aq)	→ copper carbonate (s)

[(aq) = aqueous solution; (s) = solid.]

The usual way of writing the equations would be:

$$2Ag^+(aq) + CrO_4{}^{2-}(aq) \rightarrow Ag_2CrO_4(s)$$
$$Pb^{2+}(aq) + 2I^-(aq) \rightarrow PbI_2(s)$$
$$Cu^{2+}(aq) + CO_3{}^{2-}(aq) \rightarrow CuCO_3(s)$$

2.73 *Observations on the reaction of sodium with water*

Pour a thin layer of kerosene, about 2 to 3 mm depth, on to the surface of water in a test-tube. Drop a small piece of sodium, 3-4 mm square, into the kerosene. Sodium will sink in the kerosene and float in the water. The layer of kerosene should be shallow enough to allow the top of the sodium to protrude above the surface.

The reaction between sodium and the water is much slower than if the sodium had been dropped directly on to the water. It is interesting to watch the reaction through a hand-lens held at the side (but never at the top). See the diagram.

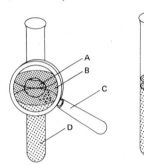

A sodium
B kerosene layer
C hand lens
D water

Observations

1. Sodium metal is lighter than water but heavier than kerosene.
2. A small area of the sodium suddenly reacts causing a stream of bubbles to appear. Is it the stream of bubbles at one side that causes movement?
3. The irregular shape of the sodium changes to that of a sphere. Has the sodium melted because the reaction gives off heat? (Melting point of sodium is 98° C.)
4. There are variations in light refraction and reflection below the sodium. Is something dissolving in the water?
5. Slight smoke where the hot sodium is above the kerosene level suggests a slight reaction with air.
6. What are the gas bubbles? Can enough of the gas be collected to show that it is hydrogen?

2.74 *Displacement of hydrogen from acids using other metals*

Pour one of the acids shown in the table below into several test-tubes to a depth of approximately 5 cm. Place a piece of different metal foil in each. Note the evolution of hydrogen and compare the different rates at which the bubbles are formed. Repeat the procedure using the other acid.

Evolution of hydrogen

Metal	3M hydrochloric acid	3M sulphuric acid
Magnesium	Very rapid	Rapid
Aluminium	Slight	None
Zinc (see note)	Moderate	Slight
Iron	Very slight	Very slight
Tin	None	None
Lead	None	None
Copper	None	None

Note.

If pupils wish to recover the zinc after the reaction has ceased, they may first obtain crystals of zinc sulphate by evaporation of the solution. Dissolve the colourless zinc sulphate crystals in water and place two carbon electrodes in the solution. Connect the electrodes to a 5- to 10-volt d.c. supply. The zinc forms rapidly on the cathode (see also experiments 2.33, 2.34).

2.75 *Preparation of sulphur dioxide*

A. A very simple method of preparing sulphur dioxide for demonstration purposes is to burn sulphur in air. This can be done by placing the sulphur in a porcelain jar, igniting it and collecting the gas formed in a funnel. The gas is then aspirated into a bottle containing water (see figure).

B. The gas can also be prepared in a generator which allows dilute sulphuric or hydrochloric

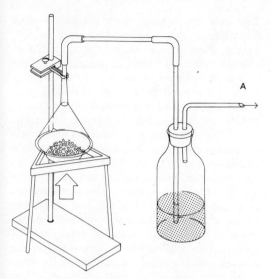

A

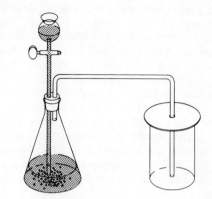

2.75B   Preparing sulphur dioxide in a generator

2.75A   Preparing sulphur dioxide by burning
            A to evacuator

acid to drip slowly on to sodium sulphite. The acid is contained in a thistle funnel and a tap controls the flow on to sodium sulphite in a suitable flask. The sulphur dioxide produced can be collected in gas jars covered with cardboard discs which have central holes for the delivery tube (see diagram B).

2.76 *Reduction using sulphur dioxide*
A. Add 10 cm³ of 0.1M solution of potassium permanganate and 10 cm³ 3M solution of dilute sulphuric acid to 200 cm³ of water containing sulphur dioxide. The solution will gradually become colourless as the sulphur dioxide reacts with the permanganate. The experiment can be continued further by stirring in a 0.25M solution of barium chloride when the solution will become 'milky' due to the formation of barium sulphate.

B. The generator used in experiment 2.75B is a convenient piece of apparatus for giving a continuous supply of sulphur dioxide for bleaching flowers and other plants. The gas from the generator is passed through a jar containing

the plant, and excess gas is absorbed in water (see the figure below). The colour of the bleached plant can easily be regenerated by placing the plant in a solution of hydrogen peroxide. This experiment could be used as an introduction to the processes of reduction and oxidation.

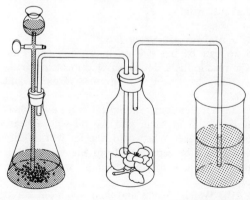

Bleaching flowers

2.77 *Reaction of magnesium with carbon dioxide*
Fill a gas-jar with carbon dioxide as described in experiment 2.38. Hold a piece of clean magnesium ribbon in a pair of tongs; ignite the magnesium with a bunsen flame and plunge it into the carbon dioxide gas. The magnesium continues to burn. Pupils might argue that the magnesium is taking oxygen from the carbon dioxide in order to burn. If this is correct, carbon will be found in the gas-jar. Are pupils able to detect carbon specks in the gas-jar? If they find this difficult, the addition of a little sulphuric acid will remove the magnesium oxide and any unburnt magnesium and make the carbon more visible.

2.78 *A simple method of titrating acids and bases*
Measure exactly 20 drops of a dilute acid such as vinegar and put these into a test-tube. Add one drop of indicator (either methyl orange or phenolphthalein is satisfactory). To this mixture of acid and indicator, add a dilute base drop by drop, and count the drops. Within experimental error, it will always take the same number of drops to neutralize the 20 drops of acid provided that the same dropper is used. A teat pipette makes a satisfactory dropper. If the concentration of the acid is known, the concentration of the base can be estimated by comparing the numbers of drops of acid and drops of base that just react.

2.79 *Making soap from fats*
Soap can be made from many oils and fats. The reaction is a double displacement involving a strong base such as sodium hydroxide and fats.
(a) Obtain some animal fat from a butcher. Boil this fat in water and the oil will separate on the surface. When cold, the fat will solidify and it can be separated from the water. Melt the fat again and strain through several layers of cloth.
(b) Weigh this fat and then weigh out about one-third as much sodium hydroxide pellets or flakes. Dissolve the sodium hydroxide in water.

Take care not to touch either the solid sodium hydroxide or the solution, because it is very caustic. Heat the fat in an iron saucepan or dish and, when it is molten, slowly add the sodium hydroxide solution with continuous stirring. Heat with a small flame to avoid boiling over. Allow the fat and the sodium hydroxide to boil for 30 minutes. Stir the mixture frequently.
(c) The next stage is to weigh out some common salt (sodium chloride), about twice the weight of sodium hydroxide used in (b) is needed. After the 30 minutes boiling, stir this salt well into the mixture. Then allow to cool. The soap separates as a layer at the top. Separate this soap from the liquid below, melt and pour into match-boxes where it will solidify again as small bars of soap. Pupils might make a comparison of the efficiency of this soap with commercially made soap.

Energy from chemical reactions

The following group of reactions involve ions in aqueous solution. When the water containing the reacting ions becomes hotter, then we have gained this heat and we can make it do work for us. During the reaction the ions have lost this heat which we have gained. On the other hand, when the water containing the ions becomes colder, it is the ions which have gained the energy and the water has lost an equivalent amount.

2.80 *Reactions which give heat energy to us*
A. Put white anhydrous copper sulphate powder to a depth of about 1 cm in a test-tube. Hold a thermometer with the bulb in the powder. Add water drop by drop. Record the changes of the thermometer reading.

B. Put about 10 cm³ of strong aqueous copper sulphate solution into a wide test-tube or small beaker. Support a thermometer with the bulb in the solution. Add magnesium powder (or ribbon) a little at a time until the blue colour disappears. What are the changes in the thermometer reading? *Caution.* The reaction is vigorous; do not carry it out in a stoppered bottle.

C. To a little water in a wide test-tube, add concentrated sulphuric acid, drop by drop, down the side of the tube. Stir gently with a thermometer after the addition of each drop. What are the changes in the thermometer reading?

2.81 *Reactions which make the surroundings colder*

Put 10 cm³ of water in a test-tube (see diagram). Read the temperature of the water. Dissolve about 2 g of potassium nitrate in the water. The temperature should fall through 9° C approximately. This means that, in the process of dissolving in the water, the particles have absorbed energy. This energy has been taken from the surrounding water in the form of heat. A similar result can be obtained by using potassium chloride instead.

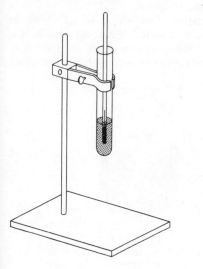

2.82 *Measuring the heat of a neutralization reaction*

Dissolve 40 g of sodium hydroxide pellets in water and make up to 500 cm³. This is a 2M solution (see Chapter One). Also prepare 500 cm³ of a 2M hydrochloric acid solution. Leave the solutions to cool to room temperature. Note the actual temperature of the solutions when cool. Then add the acid to the base quite rapidly and stir with a thermometer. Note the maximum temperature reached. The increase of temperature should be about 13° C. Since the volume of water has been doubled by adding one solution to the other, the final solution contains 1 mole of OH⁻(aq) ions which reacted with 1 mole of H⁺(aq) ions to form 1 mole of water molecules.

For the calculation we must assume that the specific heat of this moderately weak solution is the same as that of water, i.e. 1 cal per degree Celsius. Therefore the heat of neutralization, or the heat of formation of 1 mole of water molecules from the ions, is 13,000 calories or 13 kcal g-equation⁻¹. Since the reacting particles lost energy by giving this to the solution, the energy change can be written $\Delta H = -13$ kcal g-equation⁻¹ (ΔH means 'change of heat').

2.83 *Measuring the heat energy given to us by a copper displacement reaction*

Suitable reactions which are neither too vigorous nor become too hot occur between 0.2 M copper sulphate solution and iron or zinc. Using zinc, the equation may be written:

$$Zn(s) + Cu^{2+}(aq) \rightarrow Zn^{2+}(aq) + Cu(s).$$

The use of a polythene bottle for the reaction prevents some heat loss. Alternatively a glass vessel insulated with expanded polystyrene may be used.

Materials required are: polythene bottle, about 50-100 cm³, fitted with a bung and thermometer; 0.2 M aqueous copper sulphate solution; iron powder (or zinc powder) and a means of taking approximately 0.5 g samples; a measuring cylinder for 25 cm³ portions.

Put 25 cm³ aqueous copper sulphate in the

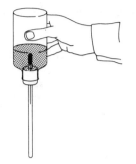

2.83 Finding the temperature rise of the reacting
        solution

bottle. Replace the bung, invert the bottle and
shake it gently (see diagram). Record the temperature of this solution. Put the bottle the right
way up, remove the bung and add about 0.5 g
of zinc dust. (This quantity is in excess and
about twice the amount required by the copper
sulphate, so that some should remain after
the reaction.) Replace the bung, invert the bottle,
and shake gently. Record the highest temperature reached. Obtain the rise of temperature.
This rise of temperature would be the same
whether 25 cm^3 or 50 cm^3 or 1,000 cm^3 of 0.2 M
copper sulphate was used. For a 1 M solution,
the rise of temperature should be multiplied
by five times. Therefore the heat of the reaction
for 1 g-formula of copper sulphate crystals is
5 × temperature increase × 1,000 calories. Since
the reacting particles lost energy, giving this to
the solution, the energy change can be written:

$$\Delta H = -5 \times \text{temperature increase},$$
$$\text{kcal g-equation}^{-1}.$$

Repeat the experiment using 0.5 g of iron powder
or iron filings. This amount is again an excess
so that all the copper sulphate will be used
up. The temperature increases obtained for a
0.2 M solution are usually about 9° to 10° C
for zinc, and 6° to 7° C for iron.

Electrical energy from chemical reactions

In the preceding experiment zinc metal became
zinc ions and copper ions became copper metal:

$$Zn(s) + Cu^{2+}(aq) \rightarrow Zn^{2+}(aq) + Cu(s).$$

This is a transfer of electrons from zinc metal
to the copper ion. In order to obtain electrical
energy these electrons must be made to flow
in an external conductor from the zinc to copper.
The potential or voltage will reflect the greater
activity of zinc over copper. The current flowing
will depend on the extent and rate of the reaction.

2.84 *Electrical energy from the displacement
       of copper by zinc*
Put some strong aqueous copper sulphate solution into a beaker. Connect copper foil to the
positive terminal of a voltmeter reading up to
5 V and a zinc rod (or foil) to the other terminal
(see figure). Dip the two metals briefly into the
copper sulphate solution. Note what changes are
shown on the voltmeter.

   The questions that will arise are: What is
the maximum reading? What happens at the
copper rod? What happens at the zinc rod? Why
does the voltage fall to zero after a short time?

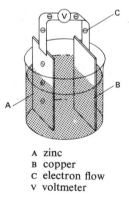

A zinc
B copper
C electron flow
V voltmeter

2.85 *Making a Daniell cell*
In the preceding experiment, copper deposited
on the zinc and caused the reaction to stop.
To prevent this happening a porous pot is used
in the Daniell cell.

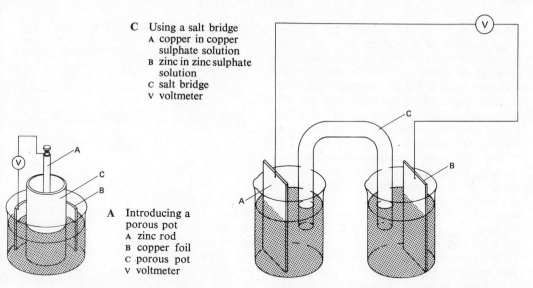

C   Using a salt bridge
A copper in copper
  sulphate solution
B zinc in zinc sulphate
  solution
c salt bridge
v voltmeter

A   Introducing a
porous pot
A zinc rod
B copper foil
c porous pot
v voltmeter

A. Put 0.5 M aqueous zinc sulphate in the porous pot. Put a strong solution of aqueous copper sulphate in the beaker surrounding the porous pot and fill to the same level as that of the zinc sulphate solution. Make a cylinder shape with copper foil and place it in the beaker to surround the porous pot (see figure). Connect the copper to the positive terminal of a 1- to 5-V voltmeter. Connect a zinc rod to the negative terminal and lower the zinc rod into the zinc sulphate solution. What reading is given by the voltmeter?

B. Insert a 1.5-V bulb in place of the voltmeter. Does it light up? Insert an ammeter into the circuit to find out the amount of current flowing. Can the current be varied by moving the copper nearer to the zinc, or by altering the surface area of the copper foil?

C. If a porous pot is not available, a salt bridge between the two solutions is almost equally effective. Make a salt bridge by filling a glass U-tube with an approximately 1 M aqueous potassium nitrate solution which has been thickened with agar. Assemble the cell as shown in the diagram and investigate the voltage, the effect on a light bulb and the current.

2.86 *Investigating an electrode potential order among the metals*

In practice, accurate values of electrode potentials of metals are derived from comparisons with the hydrogen cell under standardized conditions. Quite good comparative values can be obtained using copper and copper sulphate solution as a standard.

Filter paper, soaked but not flooded with copper sulphate solution, is laid on very clean copper foil, as shown in the diagram. A short piece of wire is clipped or soldered to the copper and connected to the positive terminal of the

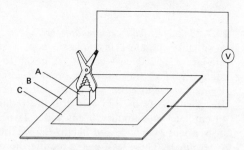

A specimen metal
B clean copper foil
c filter paper soaked in aqueous copper
  sulphate solution
v voltmeter

voltmeter (1 to 5 V). The specimen metal is gripped tightly by a crocodile clip and this is connected by a short wire to the negative terminal of the voltmeter. Clean the surface of the specimen metal and press this firmly on to the blotting paper. Record the voltage for this metal. Before testing another metal, clean the copper again with a fine emery cloth and replace the used blotting paper with a fresh piece.

If the voltmeter reading is not steady, check that: the copper surface is clean; the specimen metal surface is clean; the blotting paper contains enough copper sulphate solution; the specimen is held in good electrical contact with the crocodile clip; the specimen is held firmly on to the blotting paper.

If the voltage starts at a high value and then falls, record the highest value. The voltage drops as a deposit forms on the metal. If the voltage starts at a low value and increases, wait until the maximum value is reached. This happens with aluminium as this metal is generally covered with a film of oxide which is best removed by chemical means. Because of this oxide layer the voltage reading is initially low and increases as layer is gradually dissolved away. If aluminium is briefly dipped into strong hydrochloric acid and then pressed on to the filter paper, a better value for aluminium is obtained.

The metals magnesium, tin, lead, iron, zinc aluminium, silver can be tested by pupils. Calcium, sodium, lithium, can be demonstrated safely by the teacher.

2.87 Making a lead accumulator cell
A suitable container is a plastic cup, a small jam-jar or a 250 cm³ beaker. A cover to the container is an advantage in order to prevent drying by evaporation when the cell is not in use. Two sheets of thin lead foil are required, 40 cm long and about 10 cm wide. Two lead strips about 2 cm wide and 14 cm long are needed as terminals. These lead pieces require thorough cleaning by means of wire wool.

Fold the long sheets of lead tightly to the shorter strips so that they make good electrical contact. The projecting ends will serve as ter-

A blotting paper
B lead
C terminals

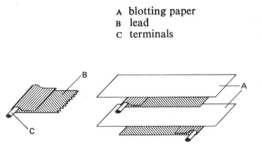

minals. A sandwich is made of alternating strips of lead foil and blotting paper (see diagram). When the sandwich is ready it is rolled up quite tightly, secured round the outside with one or two elastic bands, and placed with terminals at the top, in the cup or jar. Mark one terminal positive, and the other negative. The roll is covered with a solution of sodium sulphate made by dissolving 40 g of anhydrous sodium sulphate crystals in 200 cm³ water.

The cell is now ready to charge with electricity. This can be done with a 6-volt battery charger, or with any low voltage direct current supply giving up to 10 amps. Connect positive on the charger to positive on the cell. After only a few minutes charging, the cell will light a 1.5 volt bulb. Provided that the cell is always connected to the charger in the same way, as described above, the more times it is charged and discharged, the more efficient it becomes. There will be enough current to make a small 1 volt electric motor spin round. The cell will remain serviceable for several months if the cover is put on when not in use.

2.88 Making a dry cell
A. Set up a cell as shown in the figure using a carbon rod and zinc foil placed in a 1 M solution of ammonium chloride. The reaction is complex, but it can be thought of simply as a displacement of positive ammonium ions. Use a voltmeter to identify the positive and negative terminals. Investigate the voltage generated and the current which will flow in an external circuit. Is there enough current to light a 1.5 volt bulb?

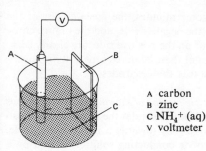

A carbon
B zinc
C NH₄⁺ (aq)
v voltmeter

B. Pour some of the ammonium chloride solution into a crystallizing dish (or any shallow dish) to a depth of about 2 cm. Add 1 cm³ of phenolphthalein indicator. Fix a carbon rod and a piece of zinc foil into crocodile clips attached to conducting wires. Join the wires together to make an electrical contact. Dip the carbon and the zinc into the solution and hold them there for several minutes. Note any changes around the electrodes. If a piece of pyrolusite (solid manganese(IV) oxide) is available, replace the carbon with this and repeat the experiment.

C. Several cells connected together are called a battery. Investigate commercial dry batteries of the type used in radios, in torches and in flash lamps for photography. The voltage given by a single cell is usually about 1.5 volts. The maximum current will vary according to the type. The currents may be investigated using an ammeter reading up to 10 amperes. Batteries for radios are designed to give small currents for long periods. An ammeter connected across the terminals of such a battery may show a current of 4 amperes whereas an ammeter connected directly across the terminals of a torch battery may show currents of 5 to 6 amperes. Flash-lamp batteries, which need to give large currents for short lengths of time, give currents larger than the above values when connected directly to an ammeter, which, in all the above cases, acts as a short circuit.

D. A 'dry' cell can be made from the following materials:

1. A zinc can; this can be zinc foil rolled into a cylinder or a cleaned-up zinc can from an old battery. Place a piece of blotting paper so that it covers the bottom of the can.
2. Carbon in very fine form; this is carbon black or acetylene black.
3. Manganese(IV) oxide, the oxidant.
4. A carbon rod, taken from an old battery.
5. Crocodile clips and leads to fix firmly to the carbon rod and to the zinc.
6. An aqueous solution of ammonium chloride. Mix 4 g of carbon black with 10 g manganese(IV) oxide. Stir in some of the ammonium chloride solution until a thick paste (like soft clay) is obtained. This takes a little time. Cut some blotting paper to make a cylinder which will go inside the zinc can. Place the carbon, manganese(IV) oxide and ammonium chloride mixture on this paper, compress it into a cylinder and wrap the blotting paper around it so that it just fits into the zinc can. When it is in the can, carefully pour a little ammonium chloride solution between the paper and the zinc, in order to ensure a good contact between the two. Press the mixture into the can firmly and tightly. Finally, fix a crocodile clip and lead to the carbon rod and press this rod into the centre of the mixture so that it does not quite touch the bottom of the can. Connect a crocodile clip and lead to the zinc container. The cell is ready. It should easily light up a 1.5-volt bulb and run a small 1.5-volt electric motor. Test the voltage and the current obtained. The carbon is one of the factors which affects the current available, as it tends to lower the internal resistance of the cell. (See also experiment 2.150.)

2.89 *Observing the movement of copper and chromate ions*

Copper chromate is a compound composed of two coloured ions, the green/blue positive copper ion and the orange negative chromate ion. The movement of the coloured ions towards electrodes may readily be observed. A 20 volt d.c. supply will be necessary. The apparatus is shown in the diagram overleaf.

Copper chromate may be prepared as follows.

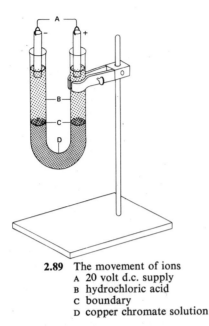

2.89 The movement of ions
A 20 volt d.c. supply
B hydrochloric acid
C boundary
D copper chromate solution

It is precipitated as a solid by adding 100 cm³ of 1 M copper sulphate solution to 100 cm³ of 1 M potassium chromate solution. The solid copper chromate is filtered quickly by using a Buchner funnel, filter flask and filter pump. It is washed with distilled water and then transferred from the Buchner funnel to a beaker where it is dissolved in the minimum quantity of bench dilute hydrochloric acid. As much urea as possible is now dissolved in the copper chromate solution in order to increase the density of the solution because it is intended that this should form a separate heavy layer below a solution of hydrochloric acid.

First fill the U-tube about one-third full of dilute hydrochloric acid. Then by means of a pipette full of the copper chromate solution, and with the jet of the pipette at the bottom of the U-tube, very carefully deliver the chromate solution so that it pushes the hydrochloric acid up and forms a separate layer below. Take out the pipette carefully to avoid mixing. The carbon electrodes should be in contact with the hydrochloric acid, and also connected to a d.c. supply of approximately 20 volts.

After some minutes, the green colour of the copper on the negative side, and the orange chromate colour on the positive side, are clearly seen. Boundaries of these coloured ions will move very slowly towards the electrodes.

2.90 *A simple method of showing movement of ions*
This experiment illustrates the movement of positive coloured ions towards a negative electrode. The electrolyte conducting solution is held by a strip of filter paper sandwiched between two microscope slides. Carbon rods serve as electrodes to lead the current through the filter paper. This is shown in the diagram. With a 10 to 20 volt d.c. supply, it is better to use the width of the slide. A greater voltage is needed to give satisfactory results if the whole length of the slide is used.

First a strip of dry filter paper is cut about 1 cm wide. A pencil mark is made across the centre of the paper. The paper is just moistened with tap water so that it is damp, but not very wet. The solution containing the coloured ion, such as Cu^{2+} or Co^{2+} ions, is placed along the pencil mark. Pupils need patience and skill to apply the ions along the mark with a fine capillary tube. The dispenser shown in the diagram might be easier to handle. A strip of filter paper 1 cm wide is folded around a thin piece of plastic material to form a

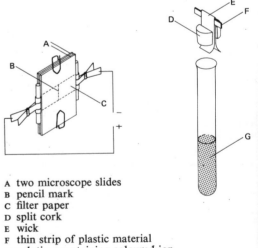

A two microscope slides
B pencil mark
C filter paper
D split cork
E wick
F thin strip of plastic material
G solution containing coloured ion

firm wick. The wick is wedged between a split cork and placed in a small test-tube containing the solution of ion. When using this dispenser, first touch the wick on to some blotting paper to remove excess solution and then lightly touch it on the pencil mark. The strip of filter paper is then sandwiched between the two slides and the ends folded round the carbon rods. The slides are held together by paper clips. The carbon rods are connected to the 20 volt d.c. supply. After some minutes the coloured ion will be seen moving towards the negative electrode. Potassium permanganate is an alternative substance to use, and it is then the coloured permanganate ion which will move towards the positive electrode.

Finding out what affects the rate of a reaction

2.91 *The smaller the particles, the greater the rate of reaction*

Marble chips can be broken up with a hammer and graded into 3 or 4 sizes: (a) coarse powder; (b) pieces about half the size of a rice grain; (c) pieces as large as rice grains; and (d) the original lumps of marble chips.

Place four 100×16 mm test-tubes in a stand. Weigh approximately 2 g of each grade (size) of marble chips and put the four grades separately into each of the four tubes. Obtain four balloons and blow them up several times to stretch them. Put 5 cm³ of bench hydrochloric acid into each of the four balloons and slip the mouth of the balloon over the top of the tube without letting any acid into the tube. This is shown in the diagram. When each balloon is in place, tip the acid into each test-tube at the same time and observe which balloon is the fastest and which is the slowest to be blown up. The smallest particles should give the carbon dioxide in the shortest time. Alternatives to marble chips for this experiment are granulated zinc, zinc foil and zinc powder, reacted with acid (*caution*: hydrogen is produced, see experiment 2.33); or aluminium foil and aluminium powder. Pupils might think of alternative reactants. Plastic bags are an alternative to balloons, but extra care is needed to fasten them to the test-tube.

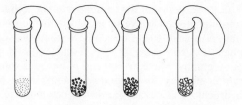

Instead of collecting the gas in a balloon or plastic bag, a more accurate method would be to collect the gas in a burette inverted over water and compare the volume of gas given off in unit time for each grade of marble chips. Another accurate method is to stand a conical flask containing the marble chips and acid on a balance and record the loss in mass every half minute. Carbon dioxide is a heavy gas and most balances will enable the loss in mass to be found as the gas escapes.

2.92 *Increased concentration of reactants increases the rate of reaction*

The reaction between sodium thiosulphate and hydrochloric acid can take a noticeable time. Sulphur is produced during the reaction making the solution cloudy. The rate of reaction can be found by finding the time taken to reach a certain degree of cloudiness in the solution. The degree of cloudiness in this case may be defined as the point at which a black cross marked below the reaction vessel can no longer be seen by looking through the solution from above. This is illustrated in the diagram.

In this experiment the concentration of sodium thiosulphate is made variable, whilst the concentration of acid is kept constant. Sodium thiosulphate may be bought as 'hypo' which is used in photography. Make up 500 cm³ of aqueous solution containing 20 g sodium thiosulphate. 2 M hydrochloric acid is also needed. Bench dilute acid is usually of this strength. Using a measuring cylinder, put 50 cm³ of thiosulphate solution into a 100 cm³ beaker. Place the beaker on a black cross marked on a sheet of paper. Add 5 cm³ of the acid and note the time given by the second hand of a clock. Stir the acid into the solution. Note the time when the cross is no longer visible through the sulphur in the solution.

Repeat the experiment with a smaller concentration of thiosulphate. Take 40 cm³ of thiosulphate solution and add 10 cm³ of distilled water. Stir and then add 5 cm³ of acid as before. The time for the cross to become invisible should be greater than for the last experiment. Repeat the experiment using 30 cm³, 20 cm³ and 10 cm³ of thiosulphate mixed with 20 cm³, 30 cm³ and 40 cm³ of distilled water.

Pupils might plot concentration of the thiosulphate solution against time taken for the reaction. Concentration values may be taken as the volume of the original thiosulphate solution used. Since 1/time (or the reciprocal of time) is the measure of the rate of the reaction, pupils might also plot thiosulphate concentrations against 1/time.

The equation for the reaction can be written as:

$$Na_2S_2O_3(aq) + 2HCl(aq) \rightarrow$$
$$\rightarrow H_2O(l) + SO_2(g) + S(s)$$

(aq = aqueous solution; l = liquid; g = gas; s = solid).

2.93 *Investigating the effect of temperature on the rate of a reaction*

The reaction in experiment 2.92 may equally well be used for investigating the effect of temperature. Start with a weaker solution. Put 10 cm³ of sodium thiosulphate solution into the 100 cm³ beaker and stir in 40 cm³ of water. Use this concentration for the series of experiments with the temperature of the solution as the variable. Add 5 cm³ of acid as

before and record the initial time and the temperature of the solution. Record the final time when the black cross below the beaker is no longer visible.

Repeat the experiment, each time warming the thiosulphate solution to just over 30° C, 40° C, 50° C and 60° C. The actual temperature of the reaction must be taken after adding the 5 cm³ of acid at the start of each experiment. The reaction goes faster at the higher temperatures. Pupils might plot a graph of temperature of reaction against time for the black cross to become invisible. As before, they might also plot a graph of temperature against 1/time.

2.94 *The effect of catalysts on the rate of reaction*

The variable in this reaction is the substance used as a catalyst in the decomposition of an aqueous solution of hydrogen peroxide. Hydrogen peroxide can usually be purchased in a chemist shop or drugstore as hair bleach.

Set up the apparatus as shown in the diagram with the burette filled with water as in a standard water displacement experiment. 2 cm³ of 20-volume hydrogen peroxide will give enough oxygen almost to fill the burette. Weigh out 1 g each

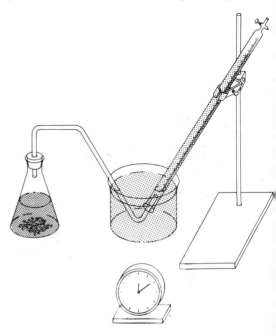

of copper(II) oxide, nickel oxide, manganese(IV) oxide and zinc oxide. Put 50 cm³ of water in the flask and add 2 cm³ of hydrogen peroxide solution. Add the 1 g of copper oxide. Immediately insert the bung with the delivery tube into the flask. Time the volume of oxygen given off at intervals of 15 seconds. Pupils might plot the volume of oxygen produced every 15 seconds against the time of the reaction.

Repeat the experiment using the other oxides as catalysts. Pupils might plot a graph for each experiment. They should now be able to discuss why manganese(IV) oxide is usually used as a catalyst in this reaction.

Pupils should try to find evidence that the catalyst is not used up during the reaction and also evidence that a catalyst may slow down a reaction as well as speed it up.

The breakdown of large molecules to small molecules

2.95 *The breakdown of starch to sugar*
Starch can be recognized by the deep blue colour which develops when it is in contact with iodine solution. This is a very sensitive test. Sugar does not react with iodine, but sugar will reduce copper(II) in Fehling's solution to red copper(I) oxide, and this is also a sensitive test. Starch does not react with Fehling's solution. Saliva contains enzyme catalysts which convert starch to sugar. This experiment investigates the progress of this reaction.

Put about 10 cm³ of dilute starch solution into a test-tube. Add to this 1 cm³ of saliva and stir this into the starch solution. Record the time of adding the saliva. At 2-minute intervals, remove 2 or 3 drops by means of a dropper and put them on a clean white tile, taking care to keep them from running into each other. The dropper must be washed well between each test. Put a little iodine solution on each drop. The decreasing intensity of the blue colour indicates that starch is being used up.

Test for increasing amounts of sugar at the same time as testing for starch. To do this, put

2 or 3 drops of the reaction mixture into a small test-tube. Add 3 cm³ of Fehling's solution and warm this mixture almost to boiling point. The test should show that the amount of sugar is increasing. The enzyme in the saliva is therefore slowly breaking starch down into sugar, which is a smaller molecule.

In a previous experiment yeast was used to break down sugar into ethanol, which is an even smaller molecule. Living yeast, which is a variety of fungus, produces enzymes which act as catalysts in the conversion:

$$\underset{\substack{\text{a simple}\\\text{sugar}}}{C_6H_{12}O_6} \xrightarrow{\text{enzyme}} \underset{\text{ethanol}}{2C_2H_5OH} + 2CO_2$$

2.96 *Breakdown of ethanol to ethene (ethylene)*
Absorb ethanol on to some cotton wool or asbestos wool and push this to the bottom of a hardglass test-tube. In the middle of the test-tube pack small pieces of unglazed porcelain. Fit a delivery tube to collect ethene gas over water as shown in the diagram. Have 3 test-tubes ready to collect ethene.

First heat the porous pot strongly and then

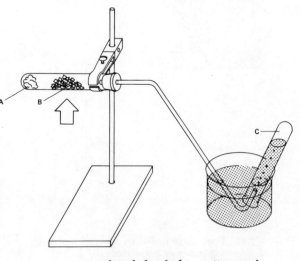

A **ethanol absorbed on cotton wool**
B **porous pot or unglazed porcelain**
C **ethene gas**

gently warm the cotton wool to produce some ethanol vapour. This vapour will break down over the hot porous pot to produce ethene gas and water vapour. The ethene is insoluble in water (unlike ethanol) and will collect in the test-tubes. Test the 3 samples by: (a) burning ethene; (b) shaking with a few drops of dilute potassium permanganate solution made alkaline with sodium carbonate solution (the colour should disappear); (c) shaking with a little bromine water (the colour again disappears). If pupils attempt this experiment they must take care to disconnect the delivery tube when they stop heating to avoid a suckback of water on to the hot porous pot.

2.97 *Breaking up a polymer into small molecules*
Experiments 2.95 and 2.96 illustrate the breakdown of solid starch to solid sugar and then to liquid ethanol and gaseous ethene. Usually the smallest molecules are gaseous or liquid at room temperature and the large molecules are solids. Perspex and polystyrene are solid polymers which can be broken down to smaller molecules by heat.

Put some pieces of perspex or polystyrene in a hard-glass test-tube. Connect a delivery tube as shown in the diagram. The collecting test-tube must be cooled thoroughly with cold water as the fumes are harmful. Gently heat the test-tube containing the perspex. The polymer will melt and give off vapours which are collected in the receiving tube. Heating must be carefully controlled to enable all the fumes to be condensed in the receiving tube. A liquid is obtained. (This suggests that the polymer has been broken down by heat to smaller molecules.) The liquid does not return to the solid state unless a catalyst is used. The specific catalysts are not usually available in school laboratories.

2.98 *Investigating the common elements in foods*
A. Collect small pieces of different foods together, such as cheese, bread, flour, sugar, leaves, maize. Heat a piece of each, about the size of a rice grain, on a tin lid or metal bottle top. Hold the lid with tongs. Which element is always left on the lid? Is it carbon?

B. Heat small amounts of food with copper oxide in a small test-tube. Copper oxide releases oxygen to the food. Test the gas in the test-tube with

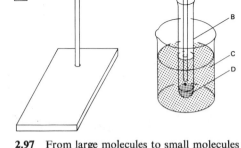

2.97 From large molecules to small molecules
  A perspex or polystyrene
  B receiving tube
  C cold water
  D a liquid collects

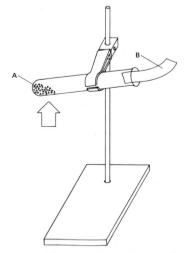

2.98C Finding nitrogen in some foods
  A food and soda lime
  B moist litmus paper

lime-water by withdrawing a little gas in a teat pipette and bubbling the gas through lime-water. Is water also condensed on the cooler parts of the tube?

C. Put a small amount of crushed food in a test-tube with 3 spatula measures of soda-lime. Mix them thoroughly and then heat as shown in the diagram. Is it possible to smell ammonia at the mouth of the tube? What colour does the litmus paper go? If the food does give off ammonia gas .in this reaction, do pupils understand that the nitrogen in the ammonia must have come from the food?

2.99 *Obtaining a gas fuel from wood*
Set up the apparatus shown in the diagram. Heat some sawdust in the hard-glass test-tube, gently at first, and then very strongly until almost red-hot. After a while the gas coming out of the jet can be ignited. Wood gas burns. The sawdust is heated without air in the test-tube. The residue is charcoal. Can pupils suggest any use for the substances condensing at A?

Building up molecules

2.100 *Extracting casein from milk*
Separate some milk from the cream, and put 100 cm³ of this milk into a beaker. Warm it to about 50° C and add acetic acid or vinegar until casein ceases to separate out. Remove the lump of casein and squeeze it with the fingers to free it from liquid. Work it in the fingers until it becomes rubbery. Casein is a protein polymer containing nitrogen atoms. If left in formalin solution it hardens. It can be moulded and made into buttons.

2.101 *A urea-formaldehyde resin*
Place 2 cm³ of 40% formaldehyde solution in a boiling-tube and add about 1 g urea. Stir until a saturated solution is obtained. Add one or two drops of concentrated sulphuric acid. The mixture suddenly hardens as it builds up to a large molecule. Extract it and wash very thoroughly. This is a condensation polymer.

2.102 *A formaldehyde-resorcinol resin*
Put 5 cm³ of 45% formaldehyde solution into a small beaker. Add 2 g resorcinol and mix very thoroughly with the formaldehyde. Add a few drops of concentrated hydrochloric acid and stir. The mixture will suddenly harden as the molecules build up into larger molecules. Extract this resin and wash very thoroughly. This is also a condensation polymer.

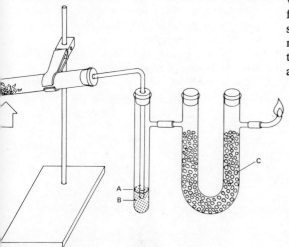

2.99 Charcoal and a gas fuel from wood
  B water to dissolve soluble gases
  C loosely packed soda lime to absorb
     water vapour

Heat and temperature

Heat as energy

2.103 *Temperature rise resulting from heat energy intake*

The amount of energy in the form of heat absorbed by different quantities of the same substance depends upon their respective masses. Place a large iron bolt and a small nail in a beaker of hot water to bring them to the same temperature. Fill two beakers with equal masses of water at the same temperature. Then place the bolt in one beaker and the nail in the other. Record the temperature of the water in each beaker after one minute. The differing amount of heat in the two objects accounts for the difference in temperature change of water in the two beakers (see diagram).

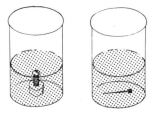

2.104 *Transfer of kinetic energy to heat energy*

A small piece of lead sheet, 5 cm² by 1 mm or thinner, is wrapped around one end of a piece of thin iron wire (25 cm long and 20 gauge, see figure). Hold the other end of the wire, with the lead resting on an anvil (an iron kilogramme weight will serve for this). Hit the lead several times in rapid succession with a hammer. Provided the lead is not more massive than suggested, the temperature will rise.

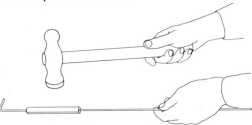

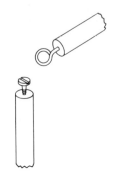

Expansion

2.105 *The ring and plug experiment*

Obtain a large wood screw and a screw eye through which the head of the screw will just pass (the eye may be fashioned out of heavy wire). Screw each part into the end of a stick, but leave at least 2.5 cm of metal protruding (see diagram). Heat the head of the screw in a flame for a while and then try to put it through the screw eye. Keep the screw hot and heat the screw eye in the flame at the same time. Now try to put the screw head through the screw eye. Keep the screw head in the flame. Cool the screw eye in cold water. Again try to put them together. Next cool the screw head and try again.

2.106 *The expansion of a solid when heated*

Obtain a piece of stout copper tubing about 2 m long. Lay it on a table and fix one end by a

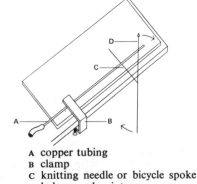

A copper tubing
B clamp
C knitting needle or bicycle spoke
D balsa wood pointer

clamp. Underneath the other end put a piece of knitting needle or bicycle spoke with a right angle bend at one end to act as a roller. A thin strip of balsa wood about 1 m long fixed to the roller by sealing wax will show any movement of the rod resting on it (see diagram). Blow steadily down the tube at the fixed end, and the expansion of the tube caused by the hot breath will be detected by this arrangement. Now pass steam through, and note the motion of the pointer. Damage to the table top can be avoided by placing a sheet of asbestos under the copper tubing. Try the experiment with different types of tubing.

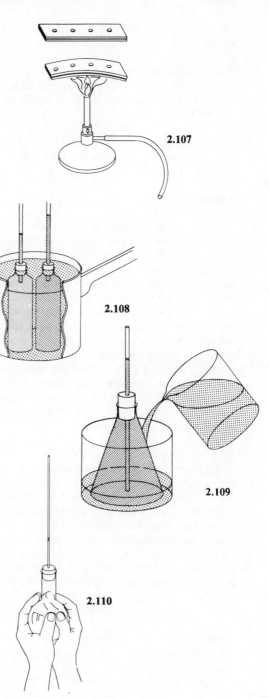

2.107

2.108

2.109

2.110

2.107 *A bimetallic strip*
A pair of iron and brass strips, riveted together, will bend when heated because of the difference of expansion. Make the holes with a nail and fix small tacks as rivets (see figure). Another way of fastening the strips together is to cut them with projections at equal intervals and bend the projections over to interlock.

2.108 *Expansion of liquids*
Fit two or three similar medicine bottles with corks and tubes. Fill them with liquids of different viscosity and immerse them in a pan of hot water (see diagram). The rise inside the tubes will indicate the difference in expansion rates.

2.109 *Expansion and contraction of a liquid*
Place some coloured water in a flask. Insert a one-hole stopper and glass tube so that it extends downward into the fluid and upward about 30 to 60 cm (see diagram). Pour warm water over the flask and the water rises in the tube. Pour cold water and the water drops inside the tube.

2.110 *Qualitative examination of expansion of air*
Air is trapped in the flask by means of a small bead of oil in the glass tubing (see diagram). Gentle heating with the hand will produce a sufficient temperature rise to cause the oil drop to move up the tubing. Then plunge the flask first into cold and then into warm (not hot) water. In

place of the flasks, glass test-tubes and corks with capillary tubing could be used.

2.111 *The expansion of air*
A. Stretch a rubber balloon over the neck of a flask. Heat the flask gently with a candle or an alcohol flame.

B. Partially inflate a balloon. Hold it over a hot plate or place it in the warm sun for a while, and observe the result.

Thermometers

2.112 *Is your temperature sense reliable?*
Fill three pans with water. Have one at the highest temperature you can bear your hand in. Fill a second one with ice-cold water. The third should be lukewarm. Put both hands in the lukewarm water and hold them there for about half a minute. Does the water seem to be the same temperature for both hands? Does it feel hot, cold or neither? Next place your left hand in the hot water and your right hand in the icy water for a minute. Quickly dry your hands and plunge both into the lukewarm water again. How does the right hand feel? Do they feel the same as when in the lukewarm water before? What do you think about your temperature sense?

2.113 *How a thermometer works*
Fill a flask with water coloured with ink. Insert a one-hole stopper carrying a 30 cm length of glass tubing until the water rises about 5 or 6 cm in the tube. Place the flask on a tripod over a heat source and observe the water level as you heat it. The water expands more rapidly than the glass and rises up the tube. Careful observation will reveal that the water level drops at the moment heating is begun, and then it begins to rise. This is because the glass bulb starts to expand before the water inside reaches the temperature of the glass.

2.114 *Making a spirit thermometer*
To make a simple alcohol thermometer accurate enough for most purposes, use 20-30 cm of glass tubing of about 5 mm external diameter with

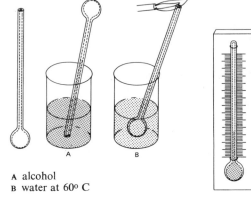

A alcohol
B water at 60° C

about 1 mm bore. A bulb of about 1.5 cm external diameter is first blown in one end of the tubing (see diagram). The tube is inverted and the open end is placed in alcohol. The bulb is alternately heated and cooled. After each cooling the alcohol drawn into the tube is shaken into the bulb. In this way the thermometer is filled with alcohol, care being taken to remove any air bubbles. The bulb of the thermometer is then placed in water at 60° C, which is slightly below the boiling point of alcohol. The excess alcohol is removed from the top of the tube as it oozes out. Now seal the open end with a hot flame. *Caution:* great care must be taken when sealing the tube. Calibrate the thermometer by placing it in water at different known temperatures.

2.115 *Testing a thermometer*
Thermometer scales are marked at two fixed points, i.e. boiling-water temperature and the temperature of melting ice. Obtain a thermometer and place it in steam immediately above the surface of water boiling in a flask. Leave it there for several minutes and notice how closely it reads 100° C or 212° F. *Note.* If you live at a high altitude, the temperature of boiling water may be well below 100° C or 212° F because of the reduced atmospheric pressure. The thermometer will read exactly 100° C only at sea level or where the barometer reading is 760 mm of mercury. Remove the thermometer from the steam, allow it to cool for

a few moments, and then place it in a jar of melting ice. Now observe how nearly it reads 0° C or 32° F.

2.116 *A simple thermoscope*
Flasks, or cut-off light bulbs, can be used to construct this apparatus (see diagram). Fit both

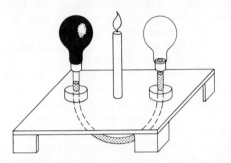

bulbs with corks and tubes about 15 cm in length. Pass the lower ends of the tubes through flat corks and, having made holes about 22 cm apart in a suitable baseboard, glue the tubes in a vertical position and connect the open ends by rubber tubing. Remove one bulb and blacken the other in a candle flame. Pour liquid into the U-tube so formed until the level is about 7 cm above the baseboard. Replace the clear bulb and slide the tube in or out a little so that the liquid remains level. Place a candle equidistant between the bulbs and wait for results.

Conduction

2.117 *How heat losses can be reduced*
Obtain four large tin cans of equal size and four smaller tin cans of equal size. Put three of the small cans inside three of the larger ones and pack insulating material under and around each of the smaller cans. Pack one with shredded newspaper, the second with sawdust and the third with ground cork (other insulating materials may be substituted if more convenient). Inside the fourth large can, place the small can resting on two corks. Fit pasteboard covers to each can. Make a hole in each cover for a thermometer. Now fill each small can

to the same depth with water that is nearly boiling. Record the temperature of the water in each can. Take the temperature of the water in each can at 5-minute intervals and notice which is the best insulator, as indicated by the slowest rate of cooling. Cooling curves may be drawn by plotting temperature versus time for each situation.

2.118 *Conduction with a metal gauze*
Hold a piece of metal gauze in an alcohol or gas flame. Observe that the flame does not go through the screen because the heat is conducted away from the flame by the wires. If you have gas in your room, place a burner under a tripod and cover it with a wire screen. Turn on the gas and light it above the screen. You will observe that the gas burns only above the screen because the screen conducts away the heat and keeps the gas below it from reaching its kindling temperature. This observation gave Sir Humphry Davy his idea for making the miners' safety lamp which does not ignite explosive gas found in coalmines.

2.119 *A model Davy lamp*
The traditional experiments on the conductivity of wire gauze can be followed by an improvised Davy lamp (see diagram). A candle enclosed in a cylinder of wire gauze does not light a jet of gas played on it from a rubber tube. A block of wood or plasticine is used as a base. *Caution.* Do not

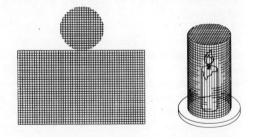

leave the gas jet turned on for extended periods. Disperse the released gas by ventilating the room.

2.120 *Snuffing out a candle flame with a copper coil*
Place a coil of heavy copper or aluminium wire over the flame of a small-sized candle (see

figure). Why does the flame go out? A candle flame can be snuffed out by depriving it of oxygen: in this case, however, the oxygen can easily get to the flame. The fire goes out because the wire conducts the heat away from the flame so fast that the temperature is lowered below the kindling point. This shows that copper and/or aluminium are good conductors of heat. If the flame is too large, it will produce heat energy too rapidly to be carried away by the coil. If the coil is already hot before the experiment, the temperature of the flame may not be lowered enough to put it out.

2.121 *Metals are good conductors of heat*
Hold a piece of paper above a candle flame: it will char if brought near. Place a metal coin on the paper and repeat the experiment: the metal will conduct the heat away and leave a pattern on the paper.

2.122 *Conduction in a metal bar*
Obtain a bar of copper, brass or aluminium at least 30 cm long. Place blobs of melted paraffin wax at 3 cm intervals. While the paraffin blobs are still soft, push the pointed ends of nails or tacks into them. Heat one end of the box with a flame. Observe the evidence that heat moves along the bar by conduction.

2.123 *Water is a poor conductor of heat*
Hold the bottom end of a test-tube of cold water in the hand. Heat the top in a bunsen flame until it boils. The fact that you can still hold the bottom shows water to be a poor conductor of heat.

Convection

2.124 *Convection in a test-tube*
Fill a hard glass test-tube with cold water. When the water is still, drop in a single, very small crystal of potassium permanganate so that it falls to the bottom, leaving little colour. First the test-tube should be held in the bare fingers, near the top of the water but not above water level. It should then be heated with a bunsen flame at the bottom of the tube as long as it can be held with bare fingers. The bunsen flame should not be too vigorous. The test-tube should be emptied, cooled, washed and filled again with cold water. When at rest, a very small crystal of potassium permanganate dye is again added without stirring. This time the tube is held at the bottom with bare fingers and heated with the bunsen flame near the top of the tube, just below the water surface. Heating should continue as long as the tube can be held. *Caution.* Remember that some children, whose skin is easily burned, do not notice much pain during the original contact. Therefore, although it is very important for children to feel the temperature changes directly, warn them to be careful not to hold the tube when it feels too hot for comfort. Test-tube holders—or their equivalent made from folded pieces of paper—spoil this experiment.

2.125 *Convection currents in water*
Fill a large jar with cold water and weigh it accurately on a balance. Empty the jar. Fill the jar with exactly the same volume of hot water and weigh. You will observe that the jar of warm water weighs less. Volume for volume, cold water is heavier than warm water; so when water is heated convection currents are set up, the warm water being lifted, because of buoyancy, by the cold surrounding water. In other words, hot water is less dense than cold, and this is the cause of convection currents in a liquid.

2.126 *Another way to show convection currents in water*
Fit an ink bottle or paste jar with a cork carrying two pieces of glass tubing, as shown in the dia-

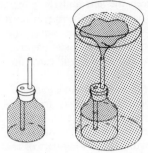

gram, but for best results one piece of tubing should be drawn out to a jet like the end of a medicine dropper. This tube should be put just through the cork and should extend about 5 cm above it. The other tube should be just level with the top of the cork and extend nearly to the bottom of the bottle. Fill the bottle with very hot water that has been coloured deeply with ink. Now fill a large glass jar such as a battery jar with very cold water. Rinse off the ink bottle and quickly place it on the bottom of the large jar. Observe what happens. Can you explain this?

2.127 *Convection currents in air*

Obtain a disc of thin tin from the end of a cylindrical tin can. Cut four blades all round the disc and pivot it on a bent knitting needle (see dia-

gram). Hold the disc above a candle flame, and it will revolve rapidly. A paper spiral supported on a knitting needle will revolve in a similar way. Another way of showing air currents is by making use of the difference in refractive index of warm and cold air. A car bulb without reflector will cast 'shadows' of convection currents from an electric heater.

2.128 *Convection currents and ventilation*

Obtain a box with grooves for a lid and cut a glass window which slides in the grooves to make an air-tight fit. Alternatively, use any box for which you can improvise a tight window. Bore four holes in each end (see diagram). Each end represents a window. The top holes of each side are the top halves of each window. Put four candles in the box, light them and close the sliding glass. You are now ready to study the best conditions for ventilation. By putting solid corks in the openings close completely both windows and observe the candles for a while. Now try different combinations of opening: one window open at the top and bottom, i.e., all four holes in one side open; one window open at the top and the other at the bottom; both windows open at the top; one open at the bottom; both open at the bottom; one open at the top. Which window openings provide the best ventilation? (See also experiment 4.120.)

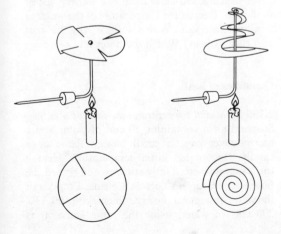

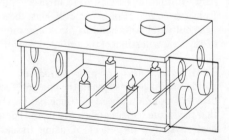

2.129 *Temperature at which water attains its maximum density*

Put a large piece of ice into a glass of water. Arrange two thermometers so that they measure the temperatures near the top and the bottom of the water. It will be noticed that the water cooled by the ice falls to the bottom; this continues until the water at the bottom of the glass reaches a temperature of about 4° C. It will stay at this temperature for a long time, the colder water remaining higher up near the ice. From this it can be deduced that the water at 4° C is denser than the water at 0° C. This curious behaviour of water is of great practical significance in nature, and explains why a pond freezes from the surface downwards while the bottom seldom falls below 4° C. (See also experiment 4.59.)

Radiation

2.130 *Heat transferred by radiation*
Heat can be transferred by wave motion, even across a vacuum. This is called radiation. Heat travels by radiation almost instantaneously. Hold your hand under an unlighted electric bulb, palm upward. Turn on the electricity. Can you feel the heat almost as soon as you turn on the bulb? The heat could not have reached your hand so quickly by conduction because air is a very poor conductor of heat. Neither could it have reached your hand by convection because this would have carried the heat upward away from your hand. It actually came to your hand carried by short electromagnetic waves of wave length longer than light. Radiation carries heat in every direction from the source. Put a piece of glass between a light bulb and your hand to block any movement of air. You will still feel the radiated warmth.

2.131 *Radiant heat waves can be focused*
Hold a reading-glass lens in the sun and focus the rays to a point on a wad of tissue paper. Observe that the tissue paper catches fire from the focused heat rays. Try the effect of using tissue paper blackened with indian ink or soot. Does it catch fire more readily?

2.132 *Radiant heat waves can be reflected*
Heat tissue paper with a reading-glass lens as above, and note the distance from the reading glass to the tissue paper. Place a tilted mirror about half way between the lens and the paper. Feel about with your hand above the mirror until you find the point where the heat waves are focused. Hold a bit of crumpled tissue paper at this point with forceps and see if it will catch fire.

2.133 *Passage of heat radiation through glass*
Hold your cheek about 25 cm away from the hole in an asbestos sheet which is fixed in front of a heating element (the sun can also be used). The hole should be level with the glowing part of the heating element. A glass plate should be inserted between the cheek and the hole. Then it should be taken out and put back, noting what you feel. The experiment may be repeated using two sheets of glass plate held together.

2.134 *Different kinds of surfaces affect radiation*
Obtain three tin cans of the same size. Paint one white, inside and out, and another black; leave the third one shiny. Fill the three cans to the same level with warm water at the same temperature. Record the temperature. Place cardboard covers with holes for thermometers on each can, set them well apart on a tray, and then put them in a cool place. Record the temperature of the water in each can at 5 minute intervals. Was there a difference in the rate of cooling? Which surface was the best radiator of heat? Which was the poorest? Empty the cans to the same level. Next fill them with very cold water, record the temperature, cover the cans and place them in a warm place or in the sun. Record the temperature of the water at 5-minute intervals. Which surface was the best absorber of heat? Which the poorest?

Quantities of heat

2.135 *Heat and temperature: the idea of a calorie*
Suspend a tin containing 50 cm³ of water and a thermometer over a small bunsen flame or a candle. Record the initial temperature. Heat it for 2 minutes, constantly stirring, and record the final temperature in degrees Celsius. Empty out the water and repeat the experiment with 100, 150, 200 cm³ of water, using the same flame. It is

sufficiently accurate to count 1 cm³ of water as 1 g. Find the product of mass of water multiplied by rise in temperature in each case. As the same heat is given out by the flame to each mass of water, the result suggests that a convenient unit of heat would be that absorbed by 1 g of water rising in temperature by 1° C. This unit is called a gramme calorie.

2.136 Caloric value for fuel
As fuels vary greatly in their heating effect, it is useful to have some way of indicating their relative effectiveness. A suitable index is the number of calories given out when 1 g of the substance

burns completely: this is called the calorific value. Hang a small can from a stand by means of wires. Pour 100 cm³ of cold water into it, and take the temperature. Place a small piece of candle on a tin lid and weigh it. Now place it under the can of water and light the wick. Stir the water with the thermometer and when the temperature reaches 60° C blow out the flame and weigh the tin lid and candle again. The mass of water (in grammes) multiplied by the rise in temperature (in °C) gives the calories produced, and the mass of candle consumed can be found from the weighings. The calorific value can be calculated from these two quantities.

Magnetism and electricity

Static electricity

2.137 Obtaining electricity by rubbing things together
Make a pile of finely divided cork particles by filing a cork. Cut up some thin paper into small pieces. Obtain a plastic comb, a plastic pencil, a plastic fountain pen, a piece of wax, a rubber balloon, a glass or china dish and any other non-metallic objects you may find. Rub each of these things briskly with your hair or a piece of fur and then bring near the pile of cork particles (see figure). Rub again and bring near the pile of thin paper. Observe what happens. Repeat the experiment, rubbing each article in turn with a silk cloth. Repeat using a piece of flannel.

2.138 Attracting water to a comb
Adjust a tap so that a very thin stream of water flows from it. Now give a comb a charge by running it through the hair several times. Hold the comb 2 or 3 cm from the stream of water. The water is strongly attracted by the charge on the comb.

2.139 The balloon stays put
Blow up a toy balloon and rub it briskly with a piece of fur. Place it against the wall and observe that it stays where you place it. Repeat, rubbing the balloon on your hair.

2.140 Repulsion with balloons
Blow up two balloons and tie with strings about one metre long. Rub the faces with fur. Hold the strings together and observe how they repel. Put your hand between them and observe what happens. Bring one of the balloons near your face. Repeat, using three balloons.

2.141 The newspaper stays on the wall
Spread out a sheet of newspaper and press it smoothly against a wall. Stroke the newspaper with a pencil or your hand all over its surface

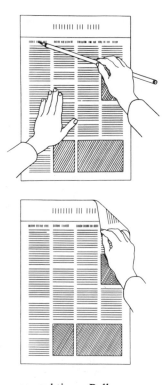

point should indent but not perforate the paper, so that the paper can turn easily. Charge a comb by rubbing on hair or wool and hold it near one end of this detector. See what happens. Try other rubbed objects.

2.143 *A pith ball indicator*
Obtain some pith from the inside of a plant stem. Dry the pith thoroughly and then press it tightly into small balls about 5 mm in diameter. Coat the pith balls with aluminium, colloidal graphite or gold paint. Attach each pith ball to a silk thread about 15 cm in length. Make a wooden stand for the pith ball (see 2.146 B). Bring objects rubbed with silk, fur or flannel near the pith ball and observe how it behaves. Such a pith ball system is called an electroscope. In place of pith balls you can use grains of puffed rice, expanded polystyrene (styrofoam) balls, ping-pong balls, or any light objects. The main thing is to make them conduct electricity by coating with metallic paint. The white of an egg should be used as an adhesive to hold powdered aluminium to the surface.

2.144 *Metal foil ball electroscope*
Roll up about 6 cm² of metal foil from a cigarette or chewing-gum packet into a ball of about 6 mm diameter. Use an adhesive to attach it to a piece of silk or nylon thread about 8 cm long. Secure the free end to a ball pen or other insulator and rest the pen across the mouth of a jam jar so that the ball hangs clear of the side (see diagram). Bring any charged body near the ball; it should first be attracted and then jump away. Now rub another plastic pen on a celluloid set-square or protractor. Hold the pen near the ball and let it take a

several times. Pull up one corner of the paper and then let it go (as shown in the diagram). Notice how it is attracted back to the wall. If the air is very dry, you may be able to hear the crackle of the static charges.

2.142 *A static electricity detector*
Cut a strip about 2 cm by 10 cm from thin cardboard. Fold it in half lengthwise and balance it on a pencil point, as shown in the diagram. The pencil

charge. Now bring the protractor near the charged ball. What does this tell you about the two kinds of charge produced by rubbing?

2.145 *A metal leaf electroscope*

To make a device for detecting charges of electricity, a jam jar, some wire, and pieces of light foil or paper are needed. A waxed cork, insulating wax or perspex are used to prevent the charge from leaking away. Push an L-shaped piece of brass or copper rod through it, and hang a folded piece of tissue paper or strip of aluminium foil from the lower end. If a charged body is brought near the rod, the leaves of the paper fly apart because they have received the same kind of charge.

2.146 *There are two kinds of static charge*

A. Make a turntable by driving a long nail through a wood base. Push a test tube into a hole made in a large flat cork. File the end of the nail to a sharp point and invert the test-tube over it. Set pins in the top surface of the cork; they will brace the objects you place on the turntable. Obtain two test-tubes or glass rods, a piece of silk, two plastic combs, an ebonite rod, some wool, and a piece of fur or flannel.

   (a) Rub a comb with fur and set it on the turntable. Rub the other comb with fur and bring it near the comb on the turntable. Repeat until you are sure your observations are correct (see illustration).

   (b) Rub a glass rod with silk and place it on the turntable. Again rub a comb with fur and bring it near the glass rod. Repeat until you are sure of your observations.

   When the comb is rubbed with fur, the plastic takes a negative charge of electricity and the fur takes a positive charge. When glass is rubbed with silk, the glass takes a positive charge and the silk a negative charge.

B. Rub an ebonite rod with a piece of wool and bring the rod near an uncharged pith ball electroscope. Observe that the pith ball is first attracted and then repelled. In a similar manner, rub a glass rod with a piece of silk and bring the rod near an uncharged pith ball electroscope. The pith ball is

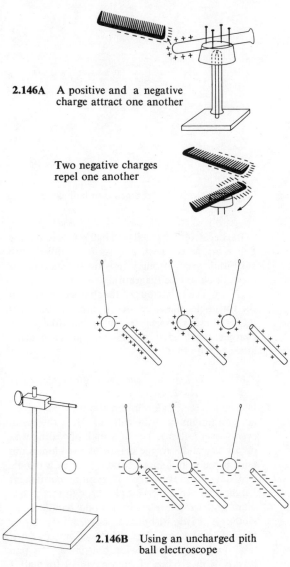

2.146A A positive and a negative charge attract one another

Two negative charges repel one another

2.146B Using an uncharged pith ball electroscope

at first attracted to the glass rod and then repelled (see figure).

C. Charge a pith ball negatively by touching it with an ebonite rod that has been rubbed with wool. Show that bringing a negatively charged rod

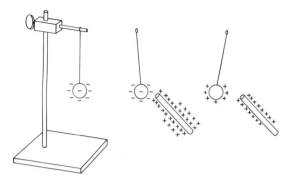

2.146C Using a charged pith ball electroscope

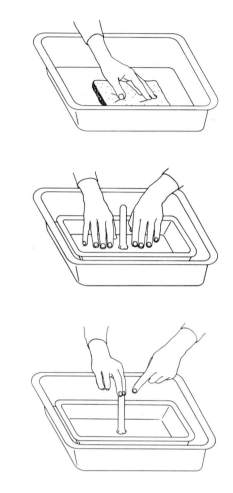

to the negatively charged pith ball repels it, but the ball is attracted when a glass rod rubbed with silk and very strongly positively charged is brought close (see diagram).

Note. The behaviour of the charged objects on the turntable, the electroscope, and the pith ball demonstrate the general rule that like static charges repel each other and unlike charges attract each other.

2.147 *Producing many charges of static electricity from one source*
Obtain a piece of aluminium about 24 cm square (an aluminium cake tin will do). Heat the metal evenly over a flame. Touch a stick of sealing wax or a wax candle to the centre of the aluminium until it melts and sticks solidly to it as a handle (see diagram). If you want a more permanent handle you can punch a hole through the aluminium and screw a plastic or wood handle to it. Obtain a plastic dishpan or bowl a little larger than the cake tin. Place the bowl or pan on a table and stroke the inside bottom of the pan briskly with a piece of fur or flannel for half a minute. Now place the aluminium on the plastic and press it down hard with your fingers. Remove the aluminium pan, place your finger near the metal and you should get a spark. You can take many charges from the plastic without further rubbing. Just press the metal against the plastic, press with your fingers and lift by the handle.

Current electricity

2.148 *A simple cell made from two coins*
Take two coins made of different metals. Clean them well with steel wool or fine sand paper. Fold some paper towelling or blotting paper into a pad so that it is slightly larger than the coins. Soak the blotting paper in salt water. Place one coin on top of the pad and the other underneath. Hold them between your thumb and finger. Connect both leads of a sensitive galvanometer to the coins and watch the deflexion.

2.149 *Electricity from a lemon*
Connect one wire from a sensitive galvanometer
to a piece of zinc cut from the can of a used dry
cell. Connect the other wire to a piece of copper.
Roll a lemon on the table, pushing on it with your
hand to break up some of the tissue inside. Push
the two metal strips through the skin of the lemon,
making sure they do not touch (see diagram).
Observe the needle.

Try this experiment using a potato. Does the
distance between the plates affect the meter reading?

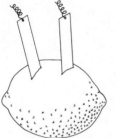

2.150 *Investigating a dry cell*
Remove the outer covering from an old dry cell.
With a saw, cut the cell in half and observe its
structure. Observe the carbon or positive pole in
the centre, the zinc can which is the negative pole
and the material between the two poles which is
the electrolyte (this is the chemical that acts on the
plates of the cell). Notice how the zinc has been
eaten away by the chemical. Observe that the
chemical materials were sealed into the zinc can
with hot pitch (see figure). (See also experiment
2.88.)

2.151 *Using a dry cell in a circuit*
Wrap the end of a short piece of stiff bell wire
around the screw base of a flashlight bulb so
that it holds the bulb tightly. Bend the remainder
of the wire in the shape of the letter C. Set the tip
of the flashlight bulb on the centre terminal of a
flashlight cell and adjust the free end so that the
springiness of the wire holds it against the bottom
of the cell (see figure). If the connexions are tight
the bulb should light. Any flashlight bulb should
operate when connected in this way, but the kind
made for a single-cell flashlight will give a much
brighter light. Look closely at the bulb and notice
the fine metal wire held in position by two heavier
wires inside. A hand lens will make this easier to
see. The fine wire is made of tungsten. Passage
of the electric current through the tungsten wire
causes it to become very hot and give off light.
Turn the cell upside down and reverse the connexions. Note that the lamp still operates, though
the electricity is flowing in the opposite direction.
Make a diagram showing the path of the current
through the bulb and around to the other end of
the cell. Develop the meaning of the term 'electric
circuit'. To use as a simple flashlight, secure wire
to the dry cell by wrapping with rubber bands.

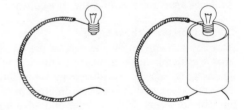

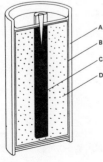

A zinc casing
B absorbent paper
C carbon rod
D electrolyte mixed with absorbent material

2.152 *A simple switch*

A simple switch can be made by fastening the end of a piece of bell wire to a pencil with two rubber bands as shown in the diagram. A second wire spliced under it makes a suitable connexion.

2.155 *Conductor or non-conductor?*

Ask the children to gather materials to be tested for electrical conductivity, and to suggest answers to this question. Try paper, eraser, plastic button, key, coins, cloth, string, chalk, glass, nail, nail

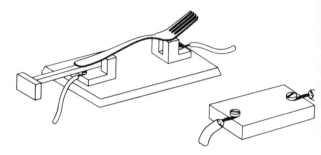

2.153 *How switches are used to control electric circuits*

Place a knife switch in a circuit with a cell and a lamp and turn the light on and off by operating the switch. Replace the lamp with a bell or buzzer and operate the switch. Replace the knife switch with a pushbutton switch. Try other common switches in the circuit. If possible, take some switches apart to show how they are constructed.

2.154 *Workings of a flashlight*

Develop recognition on the part of the children that the flashlight is an electrical device which makes use of a switch, insulators and conductors, dry cells and a bulb. Encourage children to bring in various kinds of flashlights and take them apart. Discuss the function of each part. Try to hook up the bulb to the dry cell without using the flashlight case. Reassemble the flashlight. Pupils may be challenged to find the circuit in a flashlight and to determine where the circuit is completed and broken. In metal flashlights, the case is part of the circuit. In a two-cell flashlight, the cells must be inserted so that the bottom of one cell touches the top of the other in order to provide the proper electrical circuit. Allow children to try placing the cells in various positions to discover for themselves which way works best (see figure).

file, insulated wire, bare wire, etc. Test these in a circuit across an open knife switch, or in a tester made as shown in the diagram. Materials which carry electricity are called conductors. Materials which do not carry electricity are non-conductors (insulators). The copper of a wire is a conductor; its covering is an insulator. (See also experiment 2.59.)

2.154 Workings of a flashlight

2.156 *A circuit board*

(a) All experiments using dry cells as the source of electrical energy can most easily be carried out using a simple circuit board. A piece of hardboard (or plywood) 30 cm × 30 cm is used as a base and on it are fixed clips for holding the cells, and sprung metal strips for providing connexions between cells. Brass curtain rod holders for circuit making are screwed into the base as shown in the figure.

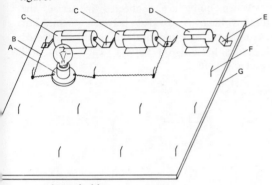

A lamp holder
B curtain wire connectors
C cells
D clips
E spring metal clips
F brass curtain rod hooks
G wooden base

(b) Spring connectors of varying lengths can be made from curtain wire with hooks inserted at each end:

(c) Lamp holders can be put into circuits by using curtain wire connectors or heavy no. 16 uninsulated copper wire:

(d) Other connexions can be made with lengths of uninsulated copper wire to the ends of which crocodile clips are attached:

2.157 *Cells in series*

A. Connect two dry cells as shown in the figure. Note that the cells are placed so that the negative terminal of one is in contact with the positive terminal of the other. When cells are connected in this way they are said to be in series.

B. Place a 4.5 volt lamp in the circuit, first with one cell, then two cells in series and finally three cells in series. Notice how the brightness of the lamp changes. When cells are connected in this way, the total voltage is the sum of the individual voltages of the cells. Therefore, if you are using 1.5 volt cells, two cells give 3 volts, and three cells give 4.5 volts.

Circuit diagram showing 2 cells in series and a lamp; the short thick line represents the negative terminal and the long thin line represents the positive terminal

2.158 *Cells in parallel*

Cells can be connected together so that all their positive terminals are joined together and similarly with all the negative terminals. They are then said to be in parallel (see figure). To investigate the behaviour of cells connected in

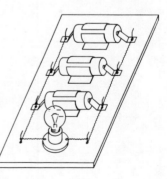

parallel, a special circuit board is necessary. If you disconnect one of the cells what happens to the brightness of the light? Is there any change when you use only one cell?

When cells are connected in parallel the total voltage is no greater than that of a single cell; however, the total current available is increased in proportion to the number of cells.

2.159 *Simple circuits*
A. *Lamps in series*

Connect up one cell and one bulb as shown in the top left-hand corner of the figure below. Note the brilliancy of the lamp. Now connect up the

other circuits as shown in the diagrams below and compare the brilliancies of the lamps as the circuits are changed. (*Note.* Make certain that all the lamps are reasonably uniform beforehand, otherwise the results will be unreliable.) These 'wiring diagrams' are useful for practical work, but pupils will need to understand theoretical circuit diagrams as they progress. Practise the use of standard symbols as shown in the equivalent circuit diagrams opposite.

When bulbs are connected in series, the total voltage is divided between them. For example, if three identical bulbs are connected in series to a 3-volt battery, each bulb will receive 1 volt.

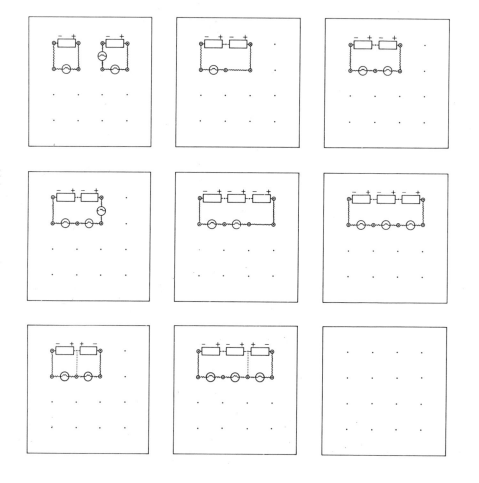

Equivalent circuit diagrams

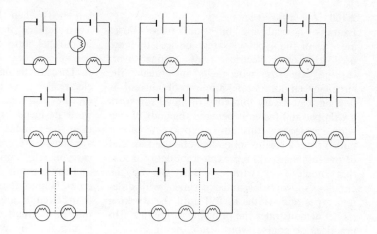

B. *Lamps in parallel*
Investigate the changes in brilliancy of the
lamps when they are connected as shown below.
When lamps are thus connected in parallel,
each one receives the full voltage of the supply.
In (d) and (e), 'chains' of bulbs (in series) are
connected in parallel across the battery. Such
an arrangement is called a series-parallel circuit.
How would you describe the circuit shown in (f)?

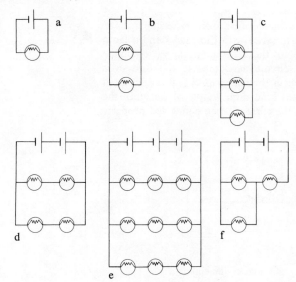

2.160 *How a fuse works*
Examine normal and burnt-out fuses. Fuses are one of the types of safety devices that are used to protect electric circuits against overloading. The fuse wire melts and breaks the circuit when an unsafe amount of current is flowing. Cut a very thin strip of metal foil from a wrapper and fasten it between the ends of two wires projecting through a cork. A thread pulled from a steel wool pad can be used instead of the foil. This will represent a model of a fuse that should work with dry cells. Set up the circuit as shown below and experiment with different types and widths of foil until the working model demonstrates the principle of the fuse. (In practice, of course, you would never connect the fuse across the supply terminals!)

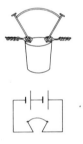

2.161 *Uses of fuses*
A. Place the model fuse from experiment 2.160 in a circuit in series with three cells and a lamp (as shown in the figure). Using a crocodile clip, short circuit the lamp. If the fuse does not melt, cut a thinner strip of foil. Experiment with different kinds and widths of foil until the foil carries the current when connected properly

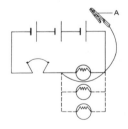

A crocodile clip

but melts when there is a 'short' in the circuit. Then replace the fuse and add more lamps in parallel until the fuse burns out.

B. Discuss the dangers of overloading domestic electrical wiring by using too many appliances on the same circuit. In older houses, circuits were designed to handle much smaller loads than they must handle today. When several appliances are used at the same time, the wires carrying the current may become overheated and cause a fire. Discuss the dangers of putting coins behind fuses and of using heavier fuses than circuits are designed for. A 30-ampere fuse in a circuit designed for a 15-ampere fuse is unsafe. To avoid accidents, only competent persons should connect up plugs to appliances.

2.162 *Getting heat and light from electricity*
Push the ends of two pieces of bell wire through a flat cork that fits a small bottle. A suitable flat cork can be made by cutting off the end of a longer cork, or a two-hole rubber stopper may be used instead. Wind the ends of a strand of very thin iron wire around the projecting ends of the copper wires and insert the cork

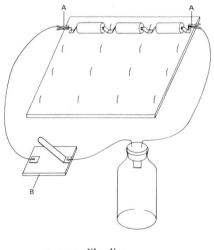

A crocodile clips
B switch

into a bottle (see diagram). The result will serve as a crude model of an electric lamp. Connect the electric lamp model into a circuit with one or more dry cells and a switch. Close the switch until the fine wire (filament) begins to glow, then open the switch again. With care the lamp can be lighted several times before the filament is consumed, but finally the heated iron wire combines with the oxygen of the air inside the bottle and burns away. Commercially made lamp bulbs contain no oxygen, and the tungsten wire is heated to such a high temperature that it glows brightly without burning. In addition to protecting the filament, the glass bulb prevents the danger of burning and electric shocks, and thus makes an electric lamp safe to use.

2.163 *Simple instruments to show electric currents*
Obtain some cotton-covered bell wire and neatly wrap from 50 to 60 turns to form a coil around a jar that is about 8 cm in diameter. Slip the coil from the jar and fasten it securely with short pieces of wire or with tape. Mount the coil on a wood base. A little platform for a compass (see diagram) can be made by cutting a hole in the cork for the coil to go through and

then fastening the cork and coil to the base with melted sealing wax or glue. Place the compass on the cork, and rotate the coil until it is in line with the needle. Connect a dry cell to the coil and observe the deflexion of the compass needle. Then reverse the connexions, and observe again. A more sensitive instrument can be made by building a little frame from cigar box wood just large enough to hold the compass. Place the compass in the frame and then wind 20 turns of bell wire over the frame as shown in the diagram.

Magnetism

2.164 *Simple compass needles*
A. Magnetize a piece of steel strip or watch spring by stroking it with lodestone or another magnet. To convert it into a compass needle, it must have as frictionless a support as possible. This can be contrived in several ways. Close a short length (2 cm) of glass tubing at the end by heating in a flame. Support this small tube on a pin pushed through a piece of wood or cork. Fix the strip of steel to the tube with glue or

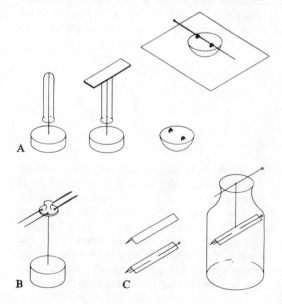

A

B   C

plasticine and adjust it so that it swings freely
and evenly (see diagram A). Another way of
supporting the compass needle is to use a metal
former from an old cloth-covered button. Clip
the magnetized rod to the two projections and
place the curved part of the button on a piece
of glass or other smooth surface.

B. Another simple compass needle can be made
using two magnetized sewing needles pushed
through the holes of a large press stud. This can
be balanced in another needle with its eye pushed
into a cork (diagram B). If a smaller press stud
is used, the flange must be squeezed between
pliers whilst pressing the needles through the
small holes.

C. Pushing a magnetized needle through cardboard and suspending it on a thread so that the
cardboard and needle balance will result in
another simple compass (see diagram C). Fix
a tiny paper arrow (with gum or shellac) to the
end of the magnet which tends to point north.

2.165 *Measuring magnetic dip angles*
Push a steel knitting needle through a cork in a
direction parallel to a diameter of its ends.
Balance it horizontally on a U-piece of brass
strip, using pins as an axle (see diagram). Take

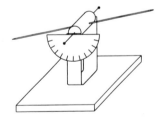

it off the knife edges and magnetize without
disturbing the cork. When it is replaced on the
bearings, one end will be pulled downwards
by the earth's magnetic field. The protractor
serves to measure this angle of 'dip'. An alternative way of suspending the magnet is to use
a piece of cycle valve tubing with a pin pushed
through it as a supporting axle. Knife edges
can be provided by two postcards held apart

by corks maintained in position by drawing pins.
The position of dip can then be marked with
a pencil and measured later.

2.166 *A magnetizing coil*
A piece of ordinary glass tubing wound with
close turns of insulated copper wire serves to
magnetize steel knitting needles. A torch battery
supplies the current required, but should not
be left connected longer than necessary (see
figure).

2.167 *Freely suspended magnet*
Suspend a magnet freely in a stirrup. This can
consist of two loops of cotton arranged to hang
the magnet as indicated in the diagram. Play
with magnets, studying the feel of attraction
and repulsion. Avoid banging the magnets to-

gether or forcing them very close together against their mutual repulsion as this might ultimately weaken them. Bring each of the poles of the magnet in your hand near the poles of the suspended magnet to investigate whether similar poles attract or repel each other (see diagram).

2.168 *Natural magnets*
Magnetic iron ore is quite common in many parts of the world. Obtain a piece of such iron ore. This is a natural magnet. Sprinkle some iron filings or finely cut pieces of steel wool on a sheet of white paper and observe how the ore attracts them. Try picking up heavier things made of iron, such as paper clips or carpet tacks. Bring the lump of ore near a compass and observe. Do all parts of the lump affect the compass in the same way?

2.169 *Obtaining artificial magnets*
Strong and useful artificial magnets for the study of magnetism can be obtained from old radio loudspeakers, from old telephone receivers and from old automobile speedometers. Magnets can frequently be purchased in the market and may be obtained from scientific supply houses. Artificial magnets are made in various shapes such as horseshoe or U-shaped and straight or bar magnets.

2.170 *Identifying magnetic substances*
Collect a variety of small objects made of paper, wax, brass, zinc, iron, steel, nickel, glass, cork, rubber, aluminium, copper, gold, silver, wood, tin, etc. Test each object with a magnet to see which ones are attracted and which are not. Bring a soft iron wire and hard steel piano wire near a compass needle to see if it is affected.

2.171 *Magnetic poles*
Break a 6 cm length from an unused steel wire and draw one end of a steel magnet along it once only and in one direction from end to end. Test the wire for magnetism with filings. Are filings attracted equally along its whole length? The areas of strongest attraction are called the poles.

2.172 *Breaking magnets*
Take a piece of magnetized steel wire with a paper arrow head on its north pole. Break the wire in half and see if you can get an isolated pole. Test both ends of each portion and make notes of the results, particularly of the sort of magnetism found on each side of the break. Break off a small piece of the wire magnet only a few millimetres in length and test it with iron filings if it is too short to test with a floating compass. What result might be expected if you could test a single grain of a powdered magnet?

2.173 *Magnetic fields (two dimensions)*
A. A bar magnet is placed on the bench. Iron filings are sprinkled on to a thin card, after which the card is placed over the magnet. The card should be supported so that it does not touch the magnet. If the card is tapped with a pencil,

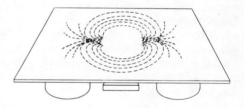

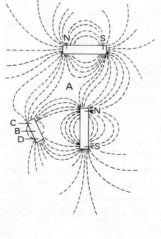

A neutral point
B soft iron
C induced S pole
D induced N pole
E screened region

the field pattern will be seen. (Iron filings can be made by shredding a steel wool pad or by filing the end of a nail held in a vice.)

B. By placing combinations of magnets on the bench, interesting fields will be observed (see figure, page 101).

You will want to preserve permanent records of your best efforts. One method is to replace the card with photographic paper (in a darkened room) and shine a bright light on it. Develop the print in the usual manner. You may also spray the arrangement of filings with black ink or varnish by means of a perfume atomizer or an air brush.

2.174 *Magnetic fields (three dimensions)*
Take a small glass jar with a tight-fitting lid. Place a spoonful of iron filings in the bottom. Add oil or other viscous liquid. Shake to see if the filings will go into suspension. If the oil is too viscous, thin with some miscible liquid until the filings remain suspended in the mixture. Placing various magnets along the sides of the jar will cause three-dimensional patterns of iron filings. A more permanent procedure is to substitute liquid plastic for the viscous solution and then let it solidify.

Electromagnetism

2.175 *Electromagnets (cylindrical)*
Obtain an iron bolt about 5 cm long which has a nut and two washers. Place a washer at each end and screw the nut just on to the bolt. Wind layers of insulated bell wire on the bolt between the washers, making certain to leave 30 cm of wire sticking out when you start winding the coil. When you have filled the bolt between the washers with several layers of turns of wire, cut the wire, again leaving about 30 cm sticking out. Twist the two ends of the wire close to the ends, then wind short lengths of tape at the ends of the bolt to keep the wire from unwinding. Remove the insulation from the two ends of wire. Connect two dry cells in series and join

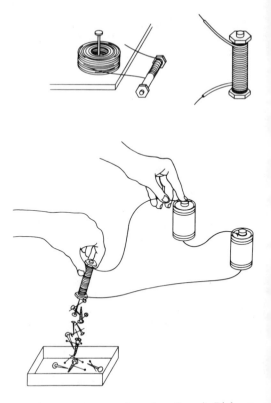

your electromagnet to them (see figure). Pick up some tacks and nails. Disconnect one wire from the battery while the tacks are still attached. Pick up other objects made of iron or steel. Test the poles at each end of the magnet with a compass while the current is turned on. Reverse the connexions to the battery. Test the poles again.

2.176 *Electromagnets (horseshoe)*
Obtain a slender bolt or a piece of iron rod about 5 mm in diameter and 30 cm long. Bend this into the shape of the letter U. Wind a coil of several layers of bell wire on each arm of the magnet, leaving the curving part free as shown. Begin at the end of one arm. Leave about 30 cm of wire sticking out for connexions. Wind about three layers on this pole, then carry the wire across the top to the other end; be sure to wind this pole exactly as shown in the diagram.

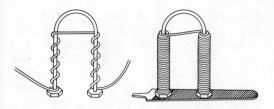

Wind about three layers of wire on this pole. When you have finished, tape the wire to keep it from unwinding. Remove the insulation from the ends of the coil, connect to two dry cells, and test the poles of the electromagnet. One should be a north pole and the other a south. If each has the same polarity, you have wound the second coil in the wrong direction. It will be necessary to unwind the coil and rewind it in the opposite direction. Try picking up different things with the magnet. Compare the strength of this electromagnet with the straight one you made.

2.177 *Testing the strength of electromagnets*
Wind 25 turns of bell wire on a straight iron bolt and connect one dry cell to the ends of this length of wire. Count the number of tacks you can pick up with the magnet. Make three trials and find the average of the number of tacks picked up. Repeat with two cells connected in series. Next, wind on 25 more turns in the same direction, join them to the first 25 turns and repeat the measurement of the strength of the magnet using first one and then two dry cells and tacks. Finally, wind on another 50 turns making 100 turns and repeat the measurement with one and two cells. As an additional experiment, remove 50 turns and rewind them on the bolt in the opposite direction. With 100 turns so wound, connect both batteries and measure the strength.

2.178 *Magnetic field due to an electric current in a wire*
A hole is made in the centre of a small white card. A length of 26-gauge copper wire about 25 cm long is put through the hole and connected to a dry cell or to the d.c. terminals of a

low-voltage supply unit. The card should be fixed in a horizontal position. Switch on the current and sprinkle iron filings on to the card. Tap the board gently with a pencil and watch the pattern develop. After this, remove the iron filings and explore with a small plotting compass. The supply connexions should be reversed to see what effect this has on the compass needle.

2.179 *Magnetic field inside an open coil*
The field inside a coil is now explored. Five spaced turns are wound on a wooden cylinder. The coil is slid off the cylinder and into slots on a card, and the ends are connected up to the d.c. terminals of a low-voltage power supply or to a dry battery (see diagram). Iron filings are sprinkled on to the card, paying particular attention to the field inside the coil. The current is switched on, the card is tapped and again the pattern is observed. Plotting compasses can also be used after the iron filings have been tried. Another name for the coil used in this experiment is an 'open solenoid'. (If the coil has many closely wound turns it is called a 'close-wound solenoid'. Using the same apparatus but with a close-wound solenoid the pupils could investigate the similarity between the field of a bar magnet and the external field of the solenoid.)

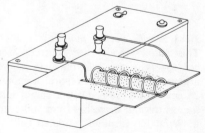

2.180 *Producing electricity with a magnet and a coil*
You will need to use one of the current detectors described in a previous section (2.163). Connect a coil of about fifty turns of bell wire to the current detector, using long connecting

wires so that the coil (and the magnet you are going to use) are kept well away from the compass in the current detector. Move the coil over one pole of a permanent horseshoe magnet, and observe the compass needle while the coil is moving through the magnetic field. Now remove the coil from the pole and observe the needle. Move the coil on and off the other pole of the magnet. Next hold the coil and plunge one pole of the magnet into the centre of the coil. Whenever the coil moves through the magnetic lines of force, a current is set up in it.

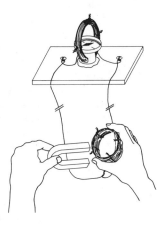

2.181 *Simple electric motor*

This simple model uses current from a dry cell to excite the field magnets as well as the armature windings. Prepare a board 20 by 25 cm for the base. Drill a small hole through the centre and drive a 15 cm spike up through it. Wind 100 turns of insulated bell wire neatly on to two other 15 cm spikes, leaving about 30 cm for free ends. Drive these spikes into the base 15.5 cm apart. Drive two small nails on the diagonal and 5 cm from the spike at the centre. Strip the free ends of each coil and twist them several times around the nails and bend them so that they will rest in contact with the central spike. These ends will serve as brushes. Care must be taken to have the

field coils wound in the proper direction. The diagram (b) is a complete plan for the direction of windings. It will work in no other way. The other ends of the coils may be fastened to the screws in the corners of the base. Your field magnets and brushes, two of the four essential parts of a motor, are now complete. The armature coil and commutator remain to be constructed. Drill a hole crosswise through the top of a 4-cm cork and force a 13-cm spike through it. Wind about 40 turns of insulated bell wire on to each end, being sure the direction of windings is as shown. Scrape the free ends. Now gouge out the centre of the cork neatly; round with a penknife and insert the closed end of a 10.5- or 13-cm test-tube so that it fits tightly. This completes the armature coil. You are now ready to make the commutator. Cut out two rectangular pieces of sheet copper about 4 cm long, and wide enough to reach around the test-tube with a gap of about 6 mm between them. Curve these to fit the tube. Punch small holes and into each solder or twist one of the scraped free ends of the armature windings. Then bind these commutator plates securely into position at top and bottom with adhesive tape. Your rotor, consisting of armature and commutator, is now complete. Set it into position on the vertical bearing and bring the brushes into contact with the commutator. Turn the test-tube in the cork until the brushes lie across the gaps in the commutator when the armature is in line with the field magnets. Now if your windings and connexions are all as shown, connect to one or two cells and with a slight push of the armature it should start off at a lively speed. If it does not go, examine the brushes to see whether they make a light, but certain contact. It may also help to change the angle of the brushes. To test this point, untwist the brushes from the nails and hold them lightly against the commutator plates with the fingers. While holding them, always parallel, swing them around at different angles while a helper turns the armature with his hand. Note the point at which the armature picks up most speed and set the brushes at that point.

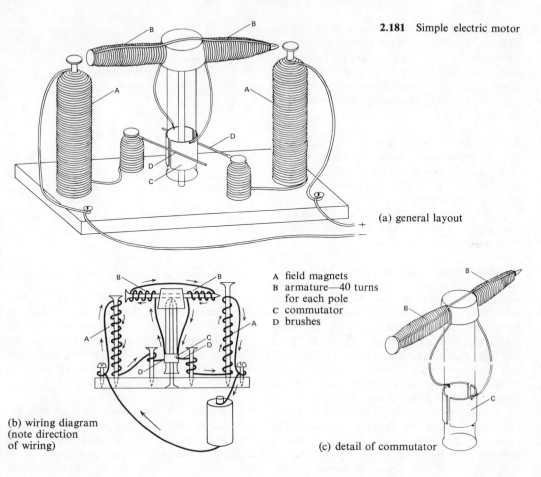

(a) general layout

A field magnets
B armature—40 turns
  for each pole
C commutator
D brushes

(b) wiring diagram
(note direction
of wiring)

(c) detail of commutator

Wave motion

Making waves

2.182 *Observing waves travelling in a rope*
Show a length of clothesline to the class. Ask
the children if they can suggest ways to make
waves travel along the rope. Have them try
their suggestions and see which ones work.
Then suggest that one child tie one end of the
clothesline to a door-knob or to a tree and pull
on the other end so that the rope does not touch
the ground. Then let him try to make large,
easily seen waves, by moving the end of the
rope up and down rhythmically to make vertical waves, or moving it to the left and right to
make horizontal waves. Get another child to
try and strike the rope rhythmically with a
stick. Do the children see the waves? (The
best place from which to observe is near one
end of the rope.) Brightly coloured pieces of
cloth tied on the rope at regular intervals will

help to make the motion more apparent to the children. Mention to them that now that they have seen the waves in the rope, the next question is 'How did the waves get there?' Encourage them to advance their theories about this. By this question the concept of applying energy may be introduced.

2.183 *Making a ripple tank*

Cut a rectangular aperture in the bottom of a photographic developing dish, size about 30 × 45 cm, leaving an edge about 2.5 cm wide all round. Fit a sheet of clear glass to the tank and stick it down to the rim, using a waterproof glue, and set it aside to dry. There are two ways of using the tank.

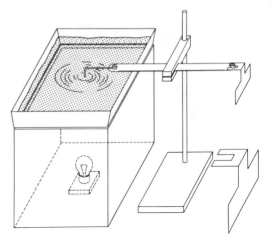

A. Obtain a cardboard box about 30 × 30 × 45 cm in size. Cut a circular hole 15 cm in diameter in the middle of one of the rectangular sides. Paint the inside of this box a dull black. As a point source of light fit a car bulb and holder to a cube of wood of 7.5 cm side. Stand the tank over the circular aperture in the box and pour in water to a depth of about 5 mm. Darken the room and switch on the bulb (see diagram). Observe the circular pattern produced on the ceiling when a drop of water falls into the water from a pen filler or pipette. If the pattern is distorted by the action of waves reflected from the sides of the tank, fit sloping 'beaches' of picture frame moulding in the water all round the edges. Should there be patterns parallel to the sides caused by the vibration of the tank as a whole, stand it on a strip of 'sorbo' rubber or on felt. Continuous trains of waves can be produced by a vibrator with one end dipping into the water. To make a vibrator clamp a 30 cm hacksaw blade in the middle and attach a single piece of stout copper wire to one end, using an electrical terminal or a small bolt. Bend the wire at right angles to the plane of the blade and cut it off about 2.5 cm long. Support the blade in a firm retort stand so that the end of the copper wire dips into the water in the tank. Pluck the free end of the blade, and notice the waves. Cut a T-shaped

piece of tin to form a dipper for plane waves and attach it to the free end of the blade as before. Stick a piece of plasticine to the blade near the copper wire so that both ends of the blade carry the same load; in this way the vibrations will be maintained for a considerable time.

B. The tank can also be set on legs with the light source over it. A large piece of white paper, or a piece of hardboard painted white, placed underneath the tank will make it easier to see the ripples. The lamp should be adjusted in height to give the best picture. The recommended depth for the water is about 5 mm. With depths below 3 mm the ripples damp out in a small distance but there is no trouble from reflections. With depths above 6 mm reflections from the edge of the tank can be troublesome. The gauze 'beaches' produce multiple, weak reflections, which may be more troublesome than a slightly stronger clear reflection produced without the gauze. If it is possible to darken the room, the ripples will be particularly effective and clear to see. If not, it is desirable to use 48-watt lamps as illuminants. Begin by telling pupils to put water into the tank to a depth of about half a centimetre and try finding out anything they can about the ripples they can make with their fingers. Suggest to pupils that the 'tartan plaid' patterns produced by jarring the tank are too complicated to yield much scientific knowledge, however pretty they are.

2.184 Simple circular pulses in ripple tanks

Start a single ripple somewhere in the middle of the tank and then make several such ripples one after the other, using: (a) a finger; (b) a pencil to touch the water; (c) a drop of water from an eye dropper.

2.185 Simple straight pulses

Pulses can be produced by giving a cylindrical wooden rod a sharp roll forward and back in the ripple tank and continuous waves are produced by continuing this motion. The ripples are rather wide near the rod but become sharper as they move away. They are sharpest when the filament of the lamp is parallel to them.

2.186 Reflection of pulses by a straight barrier

Watch what happens when a ripple (pulse) hits a wall of the ripple tank. Try it with: (a) a circular pulse; (b) a straight pulse hitting the wall 'head on' (i.e. normally); (c) a straight pulse approaching the wall in a slanting direction (i.e. incident at other angles). Avoid choosing an angle of incidence of 45° because that makes it harder to see the angle relationship. Try both a much smaller angle of incidence and a much bigger one.

2.187 Reflection at a curved barrier

Try reflecting a pulse with a curved barrier made with a rubber tube bent into something like a parabola. To help curve and weight the tube in the ripple tank, heavy copper wire should be slid into the tube before bending.

2.188 Refraction of waves

By laying a plate of glass in the middle of the ripple tank it is possible to study the transmission of waves as they pass into an apparently 'different' medium. Use a pipette to adjust the level of the water so that the plate is just covered. Notice that as the waves pass over the plate the distance between the crests (the wave length) becomes less. The velocity of the wave is also lower in the shallow water than it is in the deep water. This experiment can also be used to investigate the relation between the velocity of the waves, the wave lengths and the number of waves generated per second (frequency).

The ways in which waves are refracted depend upon the shape of the glass plate and by using plates of different shapes the refraction at a single surface, and the action of prisms and lenses can be studied.

2.189 Diffraction at narrow openings in barriers

With a gap of 2 cm or less between two barriers set up in a ripple tank, diffraction at a single slit is seen. The barriers should be about 5 cm from the vibrating beam described in experiment 2.183. At high frequencies, the waves can be seen only with a stroboscope. It will be found that waves coming round the end of the barriers are troublesome and they must be blocked off with side barriers. At high frequencies, the barriers themselves may start to vibrate giving misleading effects and this should be avoided. Alter the gap width to show less diffraction with wider apertures.

Sound

2.190 Sound wave patterns

The number of complete vibrations in one second is the frequency of a particular vibration. The way in which different sound frequencies combine is analogous to water waves. Ocean waves are longest, i.e. of low frequency. Let a small motor-boat pass over these waves. The boat sends out its own waves, which have a higher frequency than ocean waves. Next, if there is a breeze, it will send tiny ripples across the surface of the motorboat waves. The ripples usually have an even higher frequency than the other two. Now these three vibrations combine to form a pattern shown in the figure. (Fig. 2.190(a).)

In a similar way, sound waves of different frequencies from various instruments combine and form sound wave patterns. (Fig. 2.190(b).)

2.191 *Wave pattern of a tuning fork*

With a few drops of hot sealing wax attach a piece of fine wire to the prong of a tuning fork. The fork is held rigidly by the handle and placed horizontally just above the table top. Smoke a small pane of glass over the flame of an oil lamp or a candle. Now lay the smoked glass pane under the prong with the fine wire which is bent to touch the glass pane. Start the vibrations with the finger and draw the pane along the table fast enough to make a wavy line on the pane (see diagram).

Repeat this experiment drawing the pane away at different speeds and using different tuning forks.

2.192 *Seeing and feeling vibrations that make sound waves*

To demonstrate the surface vibrations that produce audible sounds, suggest the following activities:

1. Stretch and pluck rubber bands and available string instruments.
2. Hold a ruler on the edge of a desk with 15 cm extending over the edge and twang it.
3. Place a drum on a desk and scatter puffed cereal grains across the top; strike the drum and watch the cereal grains dance.
4. Press your thumb and forefinger against your larynx and make a low-pitched sound with your voice. You will be able to feel your sound vibration.
5. Hold a tuning fork loosely by the handle and strike the prongs against the edge of the desk. What do you hear? Strike the prongs again, and this time quickly touch water in a pan with the tips of the prongs. What happens? The vibrating fork will splatter the water.
6. A bell from a spoon. Cut 1 metre of cotton cord. Hold both ends together and, in the

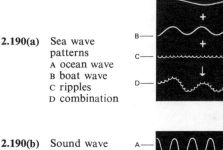

2.190(a) Sea wave patterns
A ocean wave
B boat wave
C ripples
D combination

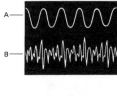

2.190(b) Sound wave patterns
A pure note
B different frequencies combine

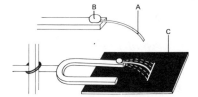

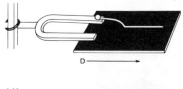

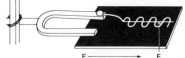

2.191 Wave pattern of a tuning fork
A fine wire
B sealing wax
C smoked glass pane on table—fork vibrating
D drawing pane—fork not vibrating
E base line
F drawing pane—fork vibrating

loop so formed, balance a teaspoon. Now press both ends into your ears with your finger tips and bend down so that the string and fork hang freely. Let someone hit the spoon lightly with a nail or another spoon. You will hear a chime like that of a bell. Sound waves travel right up the string to your ears.

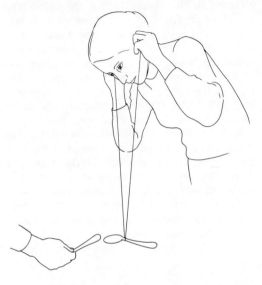

2.193 *Vibrating cans*

A. Punch a small hole in the bottom of a used tin can. Put a stout string or a fishing line through it with its end tied tightly to a pencil inside the can. Rub resin on the string. Hold the can with one hand and keep the string taut with two fingers. Now draw your fingers along the line. Sound will come from the can. Repeat the experiment of drawing your fingers along the line at different speeds. Note the different pitches of sound.

B. Two used tin cans with lids neatly cut out can be used to demonstrate a simple telephone. Punch a small hole in the bottom of each can and thread the ends of several metres of thin cotton string through the holes. Attach match-sticks or pencils to the ends of the string inside the cans. Pull the string taut and talk and listen to your pupil. Sound waves travel along the string to the bottom part of the can which acts as a diaphragm. Vibrations of the diaphragm transmit the sound waves through the air to your ear. Describe what happens when you speak into this telephone.

2.194 *Sound waves travel through wood*

To show that sound waves travel through wood, let a child rest his ear against one end of a table top. Let another child gently tap the other end of the table with a ruler or pencil.

2.195 *Testing materials that absorb sound*

Test the sound-absorbing properties of small pieces of rubber, sponge, felt, and other materials. Place the piece to be tested on a wooden table top, strike a tuning fork, and bring its handle down on a piece of material. Then strike the tuning fork again and touch its handle on the bare wood top. Which is louder? Try each material.

2.196 *Sound cannot travel through a vacuum*

To demonstrate that sound cannot travel through a vacuum it will be necessary to pump the air from a large jar or other suitable container, e.g. a Winchester bottle. If an aspirator is not available, a simple vacuum pump can be made from a bicycle pump. First open the pump and remove the piston. Unscrew the bolt that holds the leather washers and reverse the washers by turning them over. Replace the washers on the piston and reinsert the latter in the pump cylinder (see also experiment 2.309).

Place a small bell inside the jar or bottle and shake the latter while it is filled with air. You will hear the bell ringing quite clearly. Using your pump or aspirator, remove as much air as possible from the bottle and shake it again. Can you still hear the bell? How do you explain this?

2.197 *How the ear works*

Air vibrations enter the ear by the auditory passage formed at the base of the ear by the

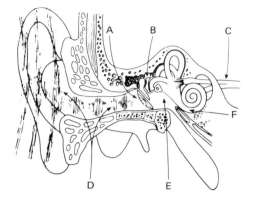

2.197 How the ear works
A ear drum vibrates
B hammer and anvil
C auditory nerve
D outer ear
E middle ear
F inner ear—semi-circular canals—
  cochlea

ear-drum membrane. They set the ear-drum in motion and, in doing so, set in motion the system of three little bones attached to it; by this means they reach a cavity in the bone called the inner ear.

One part of the ear is shaped like a snail shell. Here is found the organ which receives the sound vibrations and is connected with the brain by the auditory nerve. Another part of the inner ear which includes three small semicircular canals and serves to maintain equilibrium plays no part in hearing (see figure).

Sound vibrations are normally transmitted to the snail shell-shaped cochlea by the ear-drum and the small bones (this gives rise to a nerve message which is carried to the brain); but they can also be transmitted by the bones of the skull, and we hear a sound if the waves reach the cochlea by either route.

When a sound reaches our two ears, we can distinguish the direction from which it comes; if it comes from straight ahead, the vibrations reach both ears at the same time and with the same strength; but if the source of the sound is on one side of us, one of our ears is farther away from it and receives the waves less strongly and with a slight delay.

2.198 *How the voice is produced*
Mouth, teeth, tongue, throat and lungs are all used in the production of the voice. The sound is produced by vibrations of two thin sheets of membrane called the vocal cords, which are stretched across the sound chamber called the larynx. The larynx is the upper end of the windpipe and is located well back, at the base of the tongue. Here a trap door of cartilage called the epiglottis automatically drops down over the larynx when you swallow, so that no food will go through the windpipe (see figure). When the cords are stretched by the contraction of certain muscles in the throat, a narrow slit tends to form between them. It is when the air is forced through this narrow slit that the cords are forced to vibrate. This sets the air vibrating in the windpipe, lungs, mouth and nasal cavities.

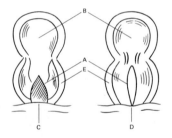

A vocal chords
B epiglottis
C ordinary breathing
D speaking
E larynx

Light

Producing light

2.199 *Suggested sources*
A compact light source can be made from any small, high intensity electric bulb which has a short, straight filament. Bulbs used in car tail lamps are excellent for this purpose. Mount the bulb on an insulating base, suitable for the voltage employed, and protect all naked terminals from accidental contact or short-circuiting. Where a very small source is needed (to make very sharp shadows) use the lamp with the filament end-on, but avoid shadow of the filament support wire. The bulbs used in direction indicators and interior car lights provide useful low-voltage line sources of light for optical experiments. A convenient holder for these low-voltage bulbs can be made from a piece of plywood. Strips of tin tacked to the wood, or held by screw terminals, can be used to make electrical connexion to the caps. A line source which operates on house voltage can be made from a clear show-case lamp. Other useful

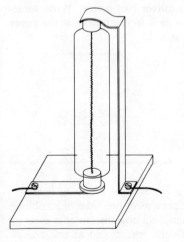

2.199 A low-voltage light source

light sources can be made from lamps designed for 35 mm slide projectors, and 8 mm movie projectors (see diagram).

2.200 *A source of light rays*
Cover the light source with a small can. Darken the room. Punch holes (1-2 mm diameter) in the can on all sides. Blow smoke around the can to make the emerging rays visible. Make enough holes so that it can be seen clearly where the light comes from and in what direction it travels.

Reflection

2.201 *Reflecting beams of light*
Hold a comb so that the sun's rays shine through the teeth and fall on a piece of white cardboard laid flat on a table. Tilt the cardboard so that the beams of light are several centimetres long. Place a mirror held upright diagonally in the path. Observe that the beams which strike the mirror are reflected at the same angle. Turn the mirror and observe how the reflected beams turn.

2.202 *Making a smoke box to study light rays*
Obtain or construct a wooden box about 30 cm wide and about 60 cm in length. Fit panes of window glass in the top and front of the box. Leave the back open, as shown in the diagram, overleaf, and cover with loosely hung black cloth that drapes like a curtain. Hang this curtain in two sections, with about a 10 cm overlap at the centre of the box. Paint the inside of the box with mat black paint. About midway between the top and bottom of one end and about 8 or 10 cm from the glass front, cut a window 10 cm high and 5 cm wide. This is to let in light rays. You can cover the window with different kinds of openings cut from cardboard and fastened with drawing pins. Cut a piece of black cardboard, make three equidistant holes about 5 mm in diameter, and fix over the window with drawing pins. Fill your box with smoke, using smouldering paper placed in a dish and

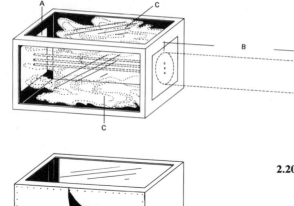

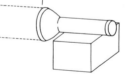

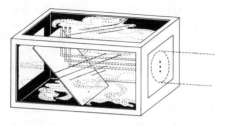

2.202 A smoke box
  A white card
  B about one metre
  C glass top and front
  D black cloth

set in one corner of the box. Next set up an electric torch or a projector about 1 metre from the window. Focus the light down to a parallel beam and direct it at the holes in the window. The light rays in the box are made visible by the smoke (see diagram).

2.203 *Regular reflection with a smoke box*
Fill the smoke box with smoke. Shine the torch beam on the three holes in the window. Now hold a plane mirror inside the box and observe how clearly the rays are still defined after reflection from the mirror. When light rays are thus reflected without scattering they are said to be regularly reflected. Move the mirror to change the angle of reflection (see figure).

2.204 *Reversed writing*
Produce reversed writing by placing a piece of carbon paper, carbon side up, under a sheet of plain paper. Write something on the paper and you will have reversed writing on the other side. Read the reversed writing by holding it in front of a mirror (see figure). Write something while you look in the mirror at the paper and watch the pencil.

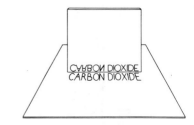

2.205 *Making a ray box for beams of light*
This apparatus consists of two sides of an oblong box 22 by 6 cm held together in this case by 2 BA rod, with the lens placed at one end of the box. The box has no bottom, and in use rests on paper pinned to a drawing board (see diagram). The light source is a 12-volt 24-watt automobile lamp. The lampholder has a sleeve

of brass tubing just fitting into a hole in a wooden slide, which forms the top of the box. The groove in front of the lens is for screens and filters. A piece of card with a slit in it provides narrow rays, and a painter's graining comb will give a bundle of rays. Convergent, parallel or divergent beams are obtained by adjusting the position of the slider. All the usual experiments with rays can be performed using slips of plane mirror, glass blocks and prisms. A curved piece of tin will show a caustic curve.

In experiments with lenses and in refraction, the lamp should be pushed down as far as possible so that the light does not pass over the top of the obstacle. A card with a hole and cross wires can be used in front of the lens as a source for optical bench experiments.

2.206 *Laws of reflection with a ray box*
A slip of mirror can be made to stand vertically by inserting one end in a piece of cork with a groove cut in it, or by means of a paperclip. Beams of light from the ray box described above are shone along the paper and marked by crosses. The incident and reflected rays, and the normal, are recorded by joining up the crosses (see diagram).

2.207 *Reflection from a concave mirror with a ray box*
Use the ray box constructed above. A concave mirror can be made from a strip of tin or a part of a metal ring. The focal length of the mirror can be measured directly by directing a parallel beam of light on to it (see diagram).

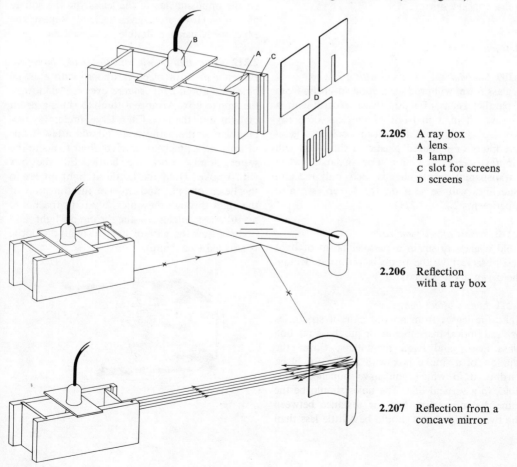

2.205   A ray box
    A lens
    B lamp
    C slot for screens
    D screens

2.206   Reflection
        with a ray box

2.207   Reflection from a
        concave mirror

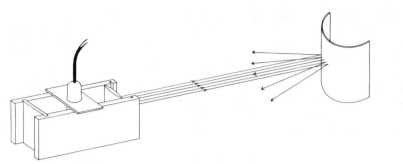

2.208 Reflection from a convex surface

2.208 *Reflection from a convex surface*
Obtain a convex mirror such as an automobile
wing mirror. Use this with the ray box and
observe the reflected rays of light (see diagram).
Compare with the reflection from a plane mirror
and a concave mirror.

Refraction

2.209 *Studying the spectrum with a ray box*
A glass prism will produce a good spectrum from
a parallel beam of light. Place a card with a
narrow slit in it in front of the lens of a ray
box (see Fig. 2.205). Interposing coloured gela-
tine filters or coloured plastic in the beam will
suppress certain colours. For instance, when
a transparent purple filter is used, only red and
blue lines will be seen on the screen (see also
experiments 2.220, 2.221).

2.210 *Prisms affect light rays*
Hold a glass prism in a parallel beam of light
and observe how the beam is refracted. Rotate
the prism about its axis.

2.211 *Lenses affect light rays*
Take the lenses from an old pair of spectacles
or used optical instruments, or purchase reading
glass lenses and hand magnifiers. Cover the
window of a smoke box with a piece of black
cardboard in which you have punched three
holes in a vertical line. The holes should be the
same distance apart, but the distance between
the two outside holes should be a little less than

the diameter of your lens. Arrange a torch to
supply parallel light rays. Fill the box with smoke
and hold a double convex lens in the path of
the three beams of light so that the middle beam
strikes the centre of the lens. Observe the beams
on the opposite side of the lens from the source
of light. How are they affected? Repeat the
experiment using a double concave lens.

2.212 *Refraction shown with the smoke box*
Fasten a piece of black cardboard with a single
hole in it about 8 mm square over the window of
the smoke box. Arrange a torch to shine a beam
of light into the box. Fill a large, preferably rectangular, bottle with water and add a few drops
of milk or a pinch of starch or flour to make the
water cloudy. Cork the bottle. Fill the box
with smoke. Hold the bottle at right angles to
the beam of light and observe the direction of
the light through the water. Next tilt the bottle
at different angles to the beam of light and
observe how the path of light through the bottle
is affected (see figure).

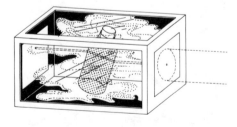

2.213 *The refraction of light and its uses*
A. Place a stick in a tall jar of water, so that part
of the stick is above the surface. Observe where
the stick enters the water and appears to be bent.
This is caused by the bending or refracting of the
light rays as they pass from water to air.

B. Place a coin in the bottom of an empty cup on
a table. Stand away, and arrange your line of
vision so that the edge of the cup just interferes
with your seeing the coin in the bottom. Hold this
position while another person pours water slowly
into the cup. What do you observe? How do you
account for this?

2.214 *Light passing through water*
A ray of light should be shown passing through a
tank of water. A converging lens is placed a suitable distance in front of the light source to produce a parallel beam of light. A screen with a
small hole (about 1 or 2 cm in diameter) placed
after the lens limits the beam to a narrow, horizontal pencil. This pencil of light is directed
through one end of the tank, which is filled with
water containing fluorescein or a small amount of
milk. The main observation is made by looking at
the pencil of light through the front of the tank.
A little smoke or chalk dust scattered in the air
would also make the beam visible in the air before
entering and after leaving the tank. Pupils may
also look through the end of the tank, looking
along the ray to see that it is straight.

2.215 *Refraction of light going from air to water*
Pour a few drops of milk into a glass of water in
order to cloud the water. Punch a small hole in a
piece of dark paper or cardboard. Place the glass
in direct sunlight, and hold the card upright in
front of the glass so that a beam of sunlight shines
through the hole. First hold the card so that the
hole is just below the water level, and observe the
direction of the beam in the water (see diagram).
Then raise the card until the beam strikes the
surface of the water (see diagram). Observe the
direction of the beam of light and experiment to
find out how the angle at which the beam strikes the
water affects the direction of the beam in the water.

2.216 *'Pouring' light*
Make a hole in the bottom of a container and fit
a stopper into it. Pour water into the container
until it is three-quarters full. Hold a flashlight in
the container so that all the light shines down
into the water. Then, in a darkened room, remove

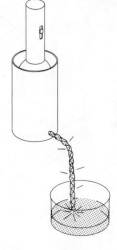

the stopper and let the water pour into the sink.
The light will appear to pour out with the water.
What happens is that some rays of light reflect
back and forth in the stream of water all the way
to the sink. Other rays are refracted to your eyes.

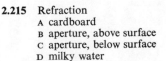

2.215   Refraction
A cardboard
B aperture, above surface
C aperture, below surface
D milky water

2.217 *How a convex lens forms an image*
Darken all the windows in a room but one. Ask a pupil to hold a lens in the window and direct it at the scene outside. Bring a piece of white paper slowly near the lens until the image picture is formed. What do you observe about the position of the image? (See figure.)

A   hand lens
B   white card
C   window

2.218 *Measuring the magnifying power of a lens*
Focus a hand lens over some lined paper. Compare the number of spaces seen outside the lens with a single space seen through the lens. The lens shown in the figure magnifies three times.

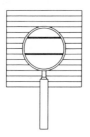

2.219 *A simple apparatus to study lenses*
All that is needed for an optical bench is a firm surface, a way of holding mirrors and lenses, and a convenient way of measuring distances (see figure).

A metre scale laid flat on the bench serves as the basis of this simple apparatus. Wooden blocks with grooves that just fit over the scale, can be adapted as holders. A layer of cork or soft cardboard glued on the top makes it easy to stick pins, such as object and search pins, into each block; strips of tin screwed to the side make convenient lens holders. A groove in the top of a block helps to keep the lens in position, and rubber tubing over the tin increases the grip.

Light sources and screens can be improvised with card and torch bulbs fastened to the blocks. It is worth while to make complete sets of this apparatus so that individual work on lenses can be attempted. The groove is easy to make with a chisel after two sawcuts have been made in the wood.

This apparatus can be used for determining the focal length of lenses (the distance from the pole of the lens of the point through which rays of light parallel to the axis of the lens are refracted) and experiments on interference and diffraction.

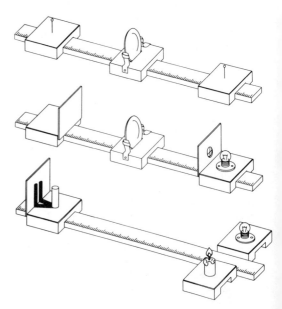

Colour

2.220 *The colour of sunlight*
Darken a room into which the sun is shining. Make a small round hole in the window shade to admit a thin beam of light. Hold a glass prism in the beam of light and observe the band of colours, called a spectrum, on the opposite wall or ceiling (see figure). Hold a reading glass lens in the colour band on the opposite side of the prism from the white sunlight. What happens to the colour band on the wall? (See also experiments 2.209, 2.221.)

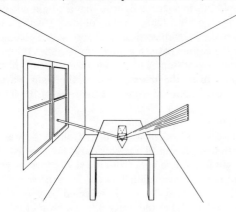

2.221 *Making a spectrum without a prism*
Set a tray of water in bright sunlight. Lean a rectangular pocket mirror against an inside edge and adjust it so that a colour band or spectrum appears on the wall.

2.222 *Colour experiments using diffraction grating material*
A very inexpensive form of this material consists of thousands of fine markings in a transparent plastic. (Diffraction grating material may be purchased from any science supply house.) The effect of these markings is to break up white light into brilliant spectra. The secret of success is to use a sharp source of light which the children view through diffraction gratings. A special light bulb with a sharp vertical line of light is placed on the desk. Each child viewing the vertical line of light sees several beautiful spectra with the colours

clearly identifiable. Pupils can easily discover for themselves the order of colours in the spectrum —ROYGBIV— representing red, orange, yellow, green, blue, indigo and violet. The same diffraction grating can be used to observe bright lines in spectra produced by fluorescent lamps and neon signs. These bright lines are characteristic of the chemical elements in the gases of the tubes and serve to identify the specific elements. This fact is the basis of the spectroscope, one of the most useful scientific instruments. (See experiment 4.101.)

2.223 *Using infra-red rays*
Heat lamps used to treat muscular ailments produce infra-red radiations, which have a wavelength longer than that of visible light. The accompanying diagram shows an effective way to produce infra-red rays and shows how they can be focused in the same manner as visible light. The iodine solution stops the visible light but allows the longer infra-red wave lengths to pass through.

The ability of the infra-red rays to pass through this solution can be related to their use in taking aerial photographs through fog and haze.

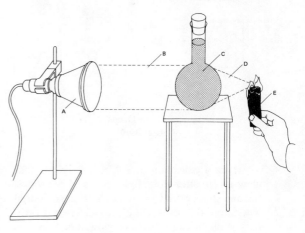

A heat lamp
B visible light
C opaque solution of iodine in carbon tetrachloride
D invisible infra-red rays
E black paper igniting

2.224 *Using ultra-violet light*
An ultra-violet light source may be used to demonstrate fluorescent phenomena. Ultra-violet sources
are available from scientific supply companies, but
a convenient apparatus for classroom demonstra-
tions can be made easily. First attach two lamp-
holders to a suitable piece of insulating material
and fasten this to the bottom of a pasteboard
carton from which the top has been removed.
Insert two ordinary, inexpensive argon lamps in
the lamp-holders (see sketch). Connect the lamps
in parallel, taking care not to leave any bare wire
exposed, and cut a notch in the side or end of the
box for the lead cord. Then turn the box upside
down and cut a peephole in what is now the top of
the box for viewing (this prevents direct eye exposure to the ultra-violet light). *Caution.* Direct
rays may cause serious damage to the eye. To
observe different objects in black light, simply
place the box over the objects and plug in the
cord.

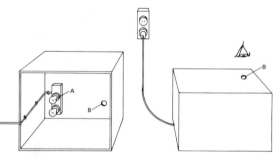

A argon lamps
B peep hole

Make a collection or exhibit of objects that
glow under ultra-violet light. Some of the socks,
ties and shirts worn by boys and girls are coloured
with fluorescent dyes. The ultra-violet rays in ordinary sunlight cause these to glow, but they will
also glow in ultra-violet light alone if placed in the
dark box under the argon lamps. Many soap powders now contain a 'brightener'. White clothes
washed in these powders will fluoresce in the ultra-
violet radiation from an argon bulb. Fluorescent
paints and lacquers are now also coming into use,

and objects coated with them may be tested and
added to the collection. Fluorescent chalk is now
commonly available and may be compared with
ordinary chalk. Certain minerals, such as wille-
mite, some fluorites, opals and sphalerites will also
fluoresce in the ultra-violet light box.

2.225 *Colours in a soap film*
Make a strong soap solution such as would be
used for blowing soap bubbles. Fill a flat dish
with the solution, and dip an egg cup or a tea cup
into the solution until a film forms across the cup.
Hold this in a strong light so that the light reflects
from the film. Observe the colours. Tilt the cup
to make the film vertical, and observe the changes
in the colour pattern as the film becomes thinner
towards the top. The colours seen in thin films
come from the interference of the light waves
reflected from the front and the back of the film.

2.226 *Colours in an oil film*
Fill a shallow dish with water. Colour the water
with black ink until it is very dark. Put the dish in
a window where light from the sky is very bright
but not in direct sunlight. Look into the water so
that light from the sky is reflected to your eye.
While looking at the water, place a drop of oil or
gasoline on the surface at the edge of the dish
nearest to you. You should see a brilliant rainbow
of colours flash away from you towards the opposite edge. By blowing on the surface you will
observe a change in the colours.

2.227 *The colour of transparent objects*
Use the smoke box constructed in experiment
2.202. Project a single ray of light into the box.
Hold a clear sheet of glass or cellophane in the

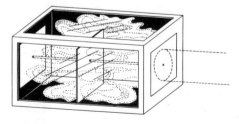

beam of light and note that the beam shining on the white screen in the box is white. Next hold a sheet of red glass or cellophane in the white beam and observe that the beam which reaches the white screen is red (see figure). All the other colours of the white light have been absorbed by the red filter. Experiment with other coloured transparent sheets. You will observe that such objects have colour due to the colours they transmit and that they absorb other colours.

2.228 *The colour of opaque objects*
Obtain a good spectrum on a wall or on a sheet of white paper in a darkened room. Place a piece of red cloth in the blue light of the spectrum. What colour is it? Place it in the green and in the yellow. How does it appear? Place it in the red light. How does it appear now? Repeat using blue, green and yellow coloured cloths. You will observe that the cloths appear black except when placed in light of the same colour. Thus opaque objects have colour because of the light they reflect; they absorb the other colours of the spectrum.

2.229 *Mixing coloured pigments*
Take a piece of blue chalk and a piece of yellow chalk. Crush them and mix them. The resulting colour is green. These are not pure colour pigments. Notice that green is between yellow and blue in the spectrum. The yellow absorbs all colours but yellow and green. The blue absorbs all colours but blue and green; thus the yellow and blue absorb each other and the green is reflected to the eye.

Try the same experiment by mixing paints from a painter's box.

2.230 *Mixing coloured lights*
The mixing of coloured lights can be achieved by using water colours painted on discs of cardboard. One suggestion is to paint a yellow 'egg yolk' on one side of a 10 cm disc, and a blue 'yolk' on the other side. When the disc is suspended between short pieces of string and twirled between the fingers and thumbs (see top figure), the result is nearly white, if the colours are carefully

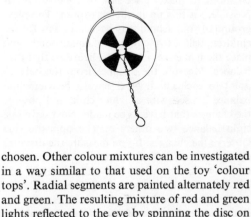

chosen. Other colour mixtures can be investigated in a way similar to that used on the toy 'colour tops'. Radial segments are painted alternately red and green. The resulting mixture of red and green lights reflected to the eye by spinning the disc on a string (see lower diagram) is yellow in this case.

2.231 *How colours change*
Paste some coloured illustrations from a magazine on a piece of cardboard. Pour three tablespoonfuls of salt in a saucer and add several tablespoonfuls of alcohol. Mix and ignite. This produces a very brilliant light that gives out only yellow. View the picture in this light in a darkened room and observe how all colours but the yellow change.

Mechanics

Balances

2.232 *Balance with a see-saw*
Obtain a strong board about 3 m long and a saw horse or box over which the board may be balanced to make a see-saw or teeter-totter. If possible, set this up in the classroom. The playground of your school may have a see-saw for the children. Select two children of equal weight and place them at either end of the board so that they balance. Measure the distance from the balance point to each child. Next have a heavier child balance himself with a lighter child and observe the changes that have to be made. Next have one child balance two others on the opposite side Observe the changes. If you measure the distance each time from the balance point to the child and multiply the distance by the child's weight you will discover an interesting thing about balance. *Note.* When two children are on the same side, measure the distance of each from the balance point, multiply by the weight of each child and add the products.

2.233 *Balance with a metre stick*
Obtain a smooth metre stick and let it rest lightly on your two forefingers. Place your fingers near the ends of the stick and then move them towards the centre. Where do your fingers meet on the metre stick? Place the finger of your right hand near one end of the metre stick and the finger of your left hand about halfway between the centre

of the stick and the other end and repeat. Where do your fingers meet this time? Now reverse and put the finger of your left hand at the end while the finger of the right hand is placed about halfway between the centre and the other end. Where do your fingers meet now?

Gravity experiments

2.234 *Ball bearings falling together*
Using two clothes pegs, a matching pair of ball bearings and a wide rubber band about 8 centimetres long, try the following experiment. Fix the band lengthwise around one of the pegs. Then open the peg and force a ball against the tension of the rubber band in between the prongs of the peg. Grip the other ball with the second peg (see the illustration). Hold the pegs side by side, pointing away horizontally above the floor. Squeeze both pegs at once. At the same moment, one ball begins to fall vertically, and the other is shot forwards. Observe what happens by looking and listening very carefully. The experiment should be repeated several times from different heights and with a tighter rubber band.

2.235 *Measuring the acceleration of marbles rolling down an incline*
Set up a grooved 3 m plank, inclined so that marbles will roll down the groove (see figure). Small tin pennants hung from simple wire axles can be arranged so that the marbles hit them and make

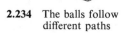

2.234 The balls follow
             different paths

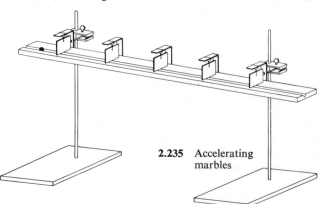

2.235 Accelerating
            marbles

'clinks'. Alternatively, the pennants could be carried by small 'goalposts' which sit over the grooved track. The goalposts could also be made from stiff pieces of wire held on either side of the track by modelling wax. Try placing the pennants at regular intervals: 25, 50, 75, 100 cm, etc., from the beginning of the plank, and try to judge or time the intervals between 'clinks'. Then try placing them so that the clinks seem to come at equal intervals of time.

2.236 *A simple pendulum*
Tie a cord at least 2 m long to some heavy object such as a stone or a small metal ball. Suspend this in a doorway or from a hook in the ceiling and start it swinging through a large arc. Count the number of swings it makes per minute. Next swing the pendulum through a short arc and determine the number of swings per minute. Repeat each of the above manipulations several times and take the average in each case. Does the length of the arc effect the time of vibration of a pendulum? Keep the length of the pendulum the same but change the material used for a weight. Repeat the manipulations suggested above. Does the material in the bob affect the time of vibration of a pendulum? Repeat each of the above experiments, but use a pendulum that is only half as long. Does the length of the pendulum affect its rate of vibration? How does it affect it?

2.237 *Coupled pendulums*
Obtain two soda-water bottles that are exactly the same size. Fill them with water and cork tightly. Place a rod across the backs of two chairs

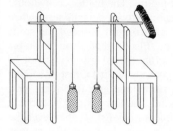

and suspend the bottles as pendulums from the rod, making sure that they are the same length (see figure). Hold one pendulum and start the other swinging; then release the first one to hang at its zero point. Soon the swinging pendulum will slow down, and the one that was quiet will take up the swing. A modification is to hang the pendulums from a stationary support such as a doorway but join the pendulum cords with a third cord tied between them about one-eighth of the way down the cords.

2.238 *Interval timing a falling body*
A. The motion of a freely falling body can be examined by attaching it to a strip of paper tape on which marks are made at equal time intervals. This may be done by passing the tape between the armature of an electric bell and a pad of carbon paper (see diagram). To modify an electric bell mechanism for this purpose, remove the clapper and extend the armature by soldering to it a strip of metal about 5 cm long. Near the end of this

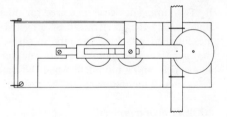

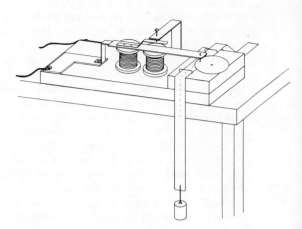

extension, drill a hole to fit a small round-headed screw, and fix it with the head downwards to act as a marking hammer. Fasten the mechanism to a piece of wood which will serve as a base. Fix another piece of wood under the striker to support the disc of carbon paper, and staples to guide the path of the ticker tape. The carbon paper disc should be about 3 cm diameter, held loosely at the centre by a drawing pin so that it can rotate to expose a new surface as the tape passes under it. The staples are easily made from wire paper fasteners pressed into the wood. The extension to the armature may have to be bent a little so that it does not strike the paper too hard and cause bouncing, which may result in uneven timing. The paper strip is now passed through the staples underneath the carbon paper and the armature is set in motion. As the strip is released and the body falls, it drags the paper after it. Marks are thus made on the paper at equal time intervals and measurements can be made of the distances travelled from the start.

B. This timing device can be used for other experiments, e.g. the acceleration of a cyclist can be measured by attaching the tape to the saddle of the machine. For more accurate measurements an a.c. bell can be modified, when the time interval is that of the frequency of the mains.

2.239 *Path of a projectile*
The apparatus shown in the diagram can be used to demonstrate that the vertical and the horizontal velocities of a projectile are independent of one another. The projectile is a metal ball, and the target is a small tin can hanging from an electromagnet. The circuit of the electromagnet includes two bared wires which are fixed parallel to and each side of the axis of a cardboard tube, and which project about 2.5 cm beyond the end of the tube (an old thermometer case with one end narrower than the other is suitable for this part of the apparatus). A large ball bearing is placed inside the tube, and is prevented from falling through by the narrow end. The electrical circuit is completed by a short length of copper wire resting on the projecting wires. Fix the tube in a

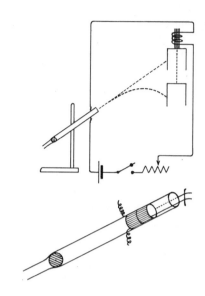

stand so that it points at the target. Blow up the tube; as the ball is propelled past the muzzle it will displace the piece of copper wire and cause the tin can to be released. The ball and target will meet in mid-air. The experiment can be repeated using different angles and distances.

Inertia

2.240 *Inertia with a stone*
You will need a stone weighing about 1 kg for this experiment. Wrap a length of heavy string around the stone. Now, on opposite sides of the stone, attach half-metre lengths of lighter cord to the heavier cord (see figure). The lighter cord should be barely strong enough to support the stone when it is suspended. Next carefully suspend the stone above a table top. (Place a board on the table under the stone so that the table top will not be dented when the rock strikes it.) Grasp the lower end of the string firmly and give it a quick jerk. If you are successful, the lower string will break and leave the stone suspended. It may even be possible to break two or three parallel strings below the stone while having just the single string above. The inertia of the stone is the cause of this.

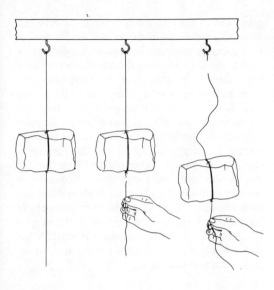

Now take hold of the remaining length of the lower string and pull steadily on it. This time the upper string breaks and the stone falls to the table because the steady application of force (rather than the quick jerk) sets the stone in motion.

2.241 *Inertia with two tin-can pendulums*
The tin cans shown in the figure must be identical; the larger they are, the more effective the demonstration. One can is suspended empty, the other is full of sand. The suspension should be as long as possible; long strings from the ceiling would be ideal. The pupils should push each can in turn to find what force is necessary to start the cans moving. They should also try stopping them when they are moving.

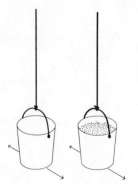

2.242 *Other inertia experiments*
A. *Inertia with a pile of books.* Stack up a pile of books. Grasp the one at the bottom of the pile. Can you remove it without upsetting the whole pile on top?

B. *Inertia with a spade.* Scoop up a spadeful of dry earth. Now pitch the earth away from you. Observe that when the spade stops the earth flies on because of inertia.

Centripetal force

2.243 *Forces with a liquid*
Obtain a small goldfish bowl or clear plastic jar. Fasten a wire securely about the neck. To this wire attach string (see figure). Clamp a hook in a hand drill chuck and attach it in the centre of the string. Place about 3 cm of water coloured with ink in the bowl. Turn the drill handle to spin the bowl and water. Observe the effects of centripetal force on the water. See if you can also observe the effects of inertia of the water when starting and stopping the bowl.

2.244 *Forces with a hard-boiled egg*
A fresh egg and a hard-boiled egg are needed for this experiment. Give each of them a spinning motion in a soup dish or a plate. You will observe that the hard-boiled egg spins longer. The inertia of the fluid contents of the fresh one brings it to

rest sooner. To understand what happens inside the egg, use a goldfish bowl as in the preceding experiment. Compare the difference in the starting and stopping behaviour using water (fresh egg) and sand (hard-boiled egg) in the bowl.

2.245 *Forces with a water bucket*
Obtain a small pail and fill it almost to the top with water. If you swing it around rapidly at arm's length in a vertical circle, the water will not spill because centripetal force acts on the water which is contained by the pail.

2.246 *Centripetal force*
Sir Isaac Newton first suggested that motion in a straight line is most natural, and that deviations from this type of motion are caused by a force pulling the body out of line. When the force acts on the body from a fixed point, such a body moves in a circle, and the force towards the centre is called a centripetal force. Circular motion can be studied by the apparatus shown in the diagram.

The force producing circular motion with different radii and frequency can be measured. Obtain a piece of glass tubing about 1 cm external diameter and cut off a 15 cm length. Heat one end in a bunsen flame until the walls of the tube are smoothly rounded. Wrap two layers of adhesive tape round the outside of the tube to provide a grip. Tie a two-holed rubber stopper to the end of about 1.5 m of nylon braided fishing line. Pass the

other end of the line down through the tube and hang half a dozen 1 cm iron washers from it. A wire paper fastener can be used as weight carrier. Adjust the line so that the distance from the top of the tube to the cork is 1 m. Grip the glass tube and swing it in a small circle above your head so that the rubber stopper moves in a horizontal circle; the force of gravity on the washers provides the horizontal force needed to keep the stopper moving in a circle. Use a small crocodile clip on the vertical fishing line to check that the motion is steady, and record the frequency of revolution required to keep the body moving in a path of radius 1 m when different numbers of washers are hung from the carrier. If the weights are doubled, how is the frequency affected? What happens when the distance from the tube to the cork is made much smaller?

Force and motion

2.247 *Effect of equal forces on light and heavy bodies*
Mark off a half-metre on a table top with chalk. Divide this into centimetres. Obtain a long rubber band and five spring clothes pegs. Attach a clothes peg to each end of the rubber band. Now grasp the clothes pegs while they rest on the table top. Place them along the marked-off place on the table top. Stretch the rubber band to a distance of

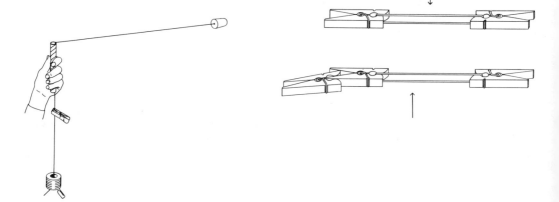

about 15 cm and release each clothes peg at the same instant. Observe that they meet half way (see figure). Next, clamp two clothes pegs on one side of the band and one on the other side. Stretch the band to a distance of about 24 cm and release. Where do they meet this time? Repeat, attaching two clothes pegs on each end of the band. Where do they meet? Again repeat with two on one side and three on the other. Where do they meet this time? Can you draw a conclusion from this experiment?

2.248 *An experiment with force and motion*
Tie a spring clothes peg open by placing one winding of thread about the long ends. Place the clothes peg in the centre of a long table and put two pencils of about equal size and weight one on either side of, and against the tied end of the clothes peg. Carefully burn the thread and observe the pencils (see figure). They are given speeds in opposite directions. Repeat the experiment, using

two larger pencils of about equal weight and size. What do you observe? Compare with the first results. Repeat, using a large and heavier pencil on one side and a small, lighter pencil on the other side. What do you observe? If you can obtain some metal balls and marbles, repeat, using different combinations of metal balls and marbles. Can you draw a conclusion from this experiment?

Action–reaction

2.249 *Action and reaction in pushing forces*
Forces work in pairs. If you push against a wall, the wall pushes back with equal force. Obtain two kitchen spring balances with square platform

tops. Put the tops together with the dial faces up. Ask a pupil to push on one while you push on the other. Observe that when you push together each balance reads the same, although the teacher may push harder than the pupil. (See also experiments 4.102, 4.103.)

2.250 *Action and reaction in pulling forces*
Obtain two spring balances. Make a loop in each end of a short piece of strong cord. Attach a spring balance to each end and ask two pupils to pull in opposite directions. Note the readings on both balances and compare them.

2.251 *Action and reaction in a model sailing boat*
Place a battery-operated fan on a sailing boat blowing against the sail. Compare results when the fan is placed on shore, blowing against the sail.

Machines

2.252 *Three types of levers*
A. Saw off a stick or board so that it is the same height as a heavy desk or table in the classroom. Place another stick of about the same length across the end of this, and use it as a lever to lift the desk or table (see figure). Note that the longer end travels farther than the shorter end. No

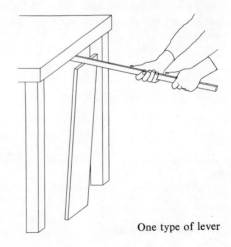

One type of lever

energy is gained, but the force exerted by the shorter end of the lever is much greater than the force used to move the longer end.

B. Obtain a uniform wooden bar about 1 m long, 4 cm wide and 5 mm thick. Drill a hole near one end, in the centre of the width dimension. Also drill holes about 12 cm from the bottom end in two uprights mounted on a base (see diagram). Place the lever bar between the uprights, and put a nail through the three holes to serve as a pivot. Place weights along the bar, and use a spring balance to measure the force applied to lift the end of the bar.

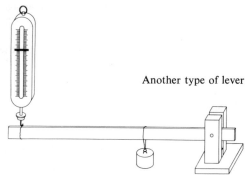

Another type of lever

C. To make a third type of lever, interchange the weight and balance (see diagram). Compare the results with those obtained above. What are the advantages and disadvantages of the three types of levers?

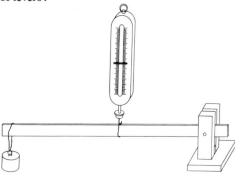

The third type of lever

2.253 *A simple wheel and axle*

Remove the cover from a pencil sharpener and tie a string tightly around the end of the shaft. Tie two or three books or a weight of several kilogrammes to the end of the string. When you turn the handle you will find the force needed to turn the handle is much less than the force of gravity on the books or weight. Point out how the pencil sharpener is being used as a wheel and axle in this experiment (see figure). Draw a diagram of the forces: does this resemble one of the types of levers in the last experiment?

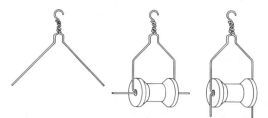

2.254 *A simple pulley*

A reasonably satisfactory pulley can be made from a wire clothes hanger and a cotton reel. Cut off both wires of the hanger at a distance of about 20 cm from the hook. Bend the ends at right angles and slip through opposite ends of the reel. Adjust the wires so that the reel turns easily and then bend the ends down to keep the wires from spreading (see sketch).

2.255 *A single fixed pulley*

Set up a single fixed pulley as shown in the diagram. By means of weights hung at A find how much force is required to lift weights of 25, 50, 75 100 and 200 g placed in turn at B. Measure the distance moved by the effort force (A) when the resistance force (weight B) is moved through 20 cm.

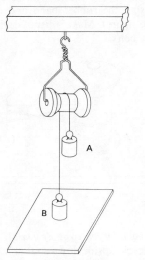

2.256 *A single movable pulley*

Suspend two pulleys on a cord from a horizontal support, and load them as shown in the diagram. If there is no adjustable support on the demonstration desk, a stick laid across the backs of two

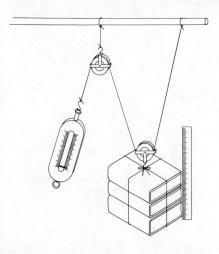

chairs can be used instead. Attach a spring balance to the end of the cord and compare the weight of the object with the force required to lift it using the pulley system. Compare also the distances through which the force and the weight are moved.

2.257 *Inclined planes*

A. Attach a heavy toy car or a roller skate to a spring balance and pull it up a sloping board (inclined plane). Note the force required to move the car and compare it with the force needed to lift it vertically. Note also that in moving up the inclined plane, the force is exerted over a greater distance than when the car is lifted vertically to the same height above the table. Neglecting friction, the work required is the same in both cases. Point out that this is also true for other simple machines (see diagram).

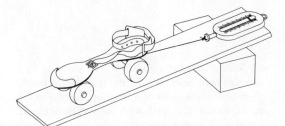

B. Mark off and cut out a right angle triangle on a piece of white paper or a piece of wrapping paper. The triangle should be about 30 cm long on its base and about 15 cm long on its shortest side. Obtain a round rod about 20 cm long and roll the triangular piece of paper on the rod beginning at the short side and rolling towards the point of the triangle. Keep the base line of the triangle even as it rolls. Observe that the inclined plane (the hypotenuse) spirals up the rod as a thread (see figure).

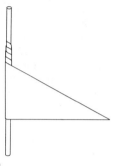

C. Bore a hole through a block of wood to fit a carriage bolt. Select a bolt that is threaded nearly its entire length. Sink the head of the bolt in the wood, so that it is flush with the surface and nail a piece of board over it. Over the projecting threads place a nut, then a washer and short piece of metal pipe. The inside diameter of the pipe must be slightly larger than the diameter of the bolt. By turning the nut with a wrench the device will act as a powerful lifting jack (see diagram).

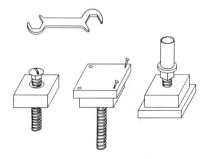

2.258 *Simple belt drives*
Drive two long nails into a block of wood. Place spools—one larger than the other— over the nails so that these can be used as axles. Slip a rubber band over both spools. Rotate the larger spool through one turn and note whether the smaller spool makes more or less than one full turn. In which direction does the small spool turn? Try crossing the rubber band and observe the result (see figure). Make a list of devices that are driven by belts.

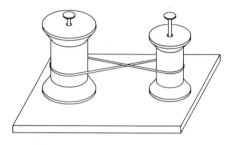

2.259 *Using a bicycle to study gears*
Turn a bicycle upside down so that it rests on the seat and handlebars. Turn the pedal wheel exactly one turn and note the number of turns made by the rear wheel.

2.260 *Simple gears*
With a hammer and a medium-sized nail, make holes exactly in the centres of several used bottle caps. Straighten the edges of the caps to make them as round as possible. Place two of the caps

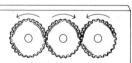

on a block of wood so that the tooth-like projections mesh together. Fasten them down with carpet tacks, but make sure that they still turn easily. Turn one of the caps and note the direction that the other turns. Add a third cap and note the direction that each turns (see sketch).

2.261 Reducing friction with pencils

Place round pencils under a heavy box. Attach a string to the box and find the force needed to move it across a table. Find the force needed to move the box without the rollers. Summarize the data obtained and suggest explanations for the results (see diagram).

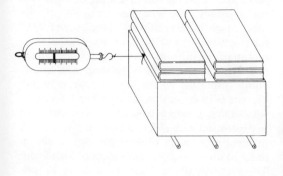

2.262 Reducing friction with wheels

Repeat the previous experiment but use a wheeled device such as a roller skate (or several roller skates) instead of rollers. State some advantages that wheels have over rollers for moving things.

2.263 Reducing friction with oil

Lay two panes of glass side by side and place a few drops of oil on one. Ask pupils to feel the difference when they rub a finger back and forth first on the unoiled pane and then on the oiled pane.

2.264 Reducing friction with ball bearings

Find two tin cans such as paint cans that have a deep groove around the top. Lay marbles in one groove and invert the other can over the marbles

to form a ball bearing. Place a book on top and note how easily the demonstration bearing turns. Oil the marbles and it will turn still more easily (see figure).

2.265 Reducing friction by an air stream

Cut out a disc of cardboard about 10 cm in diameter. With a red-hot pin, burn a hole through the centre. Saw a small cotton reel in half and glue the original end of one half over the middle of the disc. Find a piece of bamboo or some other tube which just fits the hole in the reel. Push this into the neck of a small balloon, using cotton or a rubber band to secure the joint (see figure). Blow up the balloon, pinch the neck, and insert the tube into the hole in the cotton reel. Place the disc on the table and release the air. The expanding air, escaping through the hole in the disc, will lift the card so that, given a flick, it will shoot across the table with practically no friction. This experiment illustrates the principle of the hovercraft.

2.266 *The propeller*
Although jet reaction is becoming the main pro-
pulsive force for aircraft, aviation has been well
served by the air screw, and marine propulsion is
still dependent on the screw or propeller. The
device described below can be used to exemplify
the principles involved.

The rotor can be made from the lid of a
can, diagram (a). The only provision is that the
outer edge must be rolled, to avoid any subsequent
danger from cuts. Carefully draw the three blades
on the lid. Cuts are first made along the thick
lines and then along the dotted lines so that the
smaller sections can be completely removed,
leaving three blades. The cuts are best chiselled
through on a block of wood with an old chisel.
Before giving the blades their twist, punch or drill
at the centre two 5 mm diameter holes 5 mm
apart, then remove the little bridge of metal
between them to make a central slot.

The next requirement is a twisted strip of thick
metal about 1 cm by 25 cm to fit the above slot.
As this may not be readily available you may
prefer to use two strong wires, see diagram (b).
To twist the wire, fold a 60 cm length in half with
a large loop at the bend. Slip rod B through the
loop and clamp the free ends close together in a
vice. Then twist up the doubled piece to give a
long uniform twist of angle about 20° to the axis.
The holes in the rotor may need a little trimming
so that it will spin freely up and down the twist.

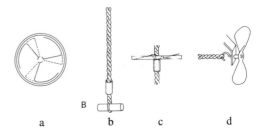

a   b   c   d

Finally, obtain a short tube that slides easily along
the wire. Tin may be rolled up for this purpose.
The angle of the rotor blades must naturally be
such as to give lift when the rotor is spun by
being pushed off the wire. There are three parts to
this assembly. The wire, held vertically; the tin
tube, which should rest on the loop at the foot of
the twisted wire; and the rotor, which should rest
on top of the tin tube, as shown (c). To launch
this flying saucer, hold the arrangement steady
above your head by the tube and strongly pull
the wire twist down with the other hand. Since the
whole thing is quickly constructed, you may wish
to experiment with different blade angles, or with
different numbers of blades from two to six, in
order to obtain the optimum effect of high flight.
These tests should be made out of doors. Propel-
lers of the form shown in the diagram (d) are
easily cut from sheet metal. With a rubber-band
drive they are a familiar source of power both on
model aeroplanes and model boats.

Fluids

Liquid pressure

2.267 *The difference between weight and pressure*
Make two square blocks of wood, one much smaller than the other, and join them together as
shown in the diagram. Press each of these faces
consecutively into a slab of clay or plasticine using
the same force in each case. The difference in
pressure is seen by the different depths of the
indentations (see figure).

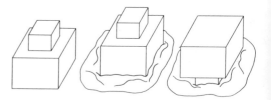

2.268 *Liquids exert pressure*
Connect two 15 cm lengths of glass tubing or two transparent plastic soda straws with a short length of rubber tubing and attach them to an upright board as shown in the diagram. Put some coloured water in the tubes to a depth of about 6 or 8 cm. This is your pressure gauge or manometer. Stretch a piece of thin rubber tightly over a small funnel and tie it securely with thread or string. Attach the funnel to the manometer with a 30 cm length of rubber tubing. Push the funnel into a pail of water and watch the manometer liquid levels change.

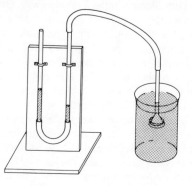

2.269 *Water pressure changes with depth*
Fill a tall glass jar or pail with water. Using the funnel and manometer you made in the previous experiment, measure the pressure just below the surface with your manometer. Measure the pressure at the bottom. How does pressure change with depth?

2.270 *Pressure depends upon the liquid*
Obtain two glass jars into which the funnel used in the two previous experiments will fit. Pour water into one jar and a less dense liquid such as alcohol into the other. Make sure the depths of the liquids are the same. Measure the pressure at the bottom of the jar of water. Measure the pressure at the bottom of the jar of alcohol. How do they compare?

2.271 *Water pressure in a large vessel*
Use the funnel and manometer from the preceding experiments. Obtain a small glass jar of small diameter and a glass jar of larger diameter. Fill both jars to the same depth with water. Measure the pressure at the bottom of each jar. How do they compare?

2.272 *Water pressure is the same in all directions*
Punch holes around the base of a tall tin can with a nail. Cover the holes with a strip of tape. Fill the can with water and hold it over a sink. Strip off the tape. Observe and compare the distance the streams shoot out from the holes all around the can.

2.273 *Balancing water columns*
Drill a hole in or remove the bottoms from several plastic bottles of different shapes but of about the same height. Fit the bottles with stoppers or corks carrying glass tubes as shown in the diagram. Connect the bottles together as shown. Pour coloured water into the bottles until they are nearly full. This experiment shows that in a given liquid, pressure is independent of the size or shape of the vessel and depends only on the depth.

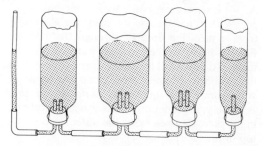

2.274 *Raising heavy weights by water pressure*
Obtain a rubber hot-water bottle. Put a one-hole stopper carrying a short glass tube tightly in the neck. Punch a hole in the bottom of a tin can and make it large enough to take a one-hole stopper. Put a short length of glass tube through the stopper. Connect the water bottle and the can with a length of rubber tube at least 1.25 metres in length—it will be wise to wind wire around the

connexion at the bottle. Fill the bottle, tube and can with water. Place the bottle on the floor and put a length of board on it. Place books or other heavy objects on the board (see figure). Now raise the can above the level of the floor and observe what happens to the books. See how heavy a weight you can lift by raising the can as high above the floor as possible.

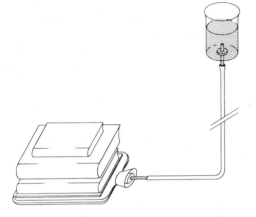

2.275 Water will not compress
A. Fit a soda-water bottle with a one-hole stopper. With its narrow end up, place the glass tube from a medicine dropper through the stopper. Fill the bottle to the top with water. Insert the stopper tightly until the water rises a little way in the medicine dropper. Now grasp the bottle in your hands and squeeze as hard as you can. Because you cannot compress water, it will rise in the tube. Can you make the water run over the top of the tube?

B. Fill a medicine bottle with water. Force in a good cork. Strike the cork sharply with a hammer; the bottle will burst. *Caution:* Wrap material round the bottle to avoid the danger of flying glass.

2.276 Making a model hydraulic lift
Some freight and passenger lifts are raised by water pressure. You can make a model of one of these with an automobile hand pump. Connect the tube from the pump to one end of a length of rubber hose. Bind the connexions with wire so that they will not blow out. Now connect the other end of the hose to a water tap, again binding the connexion with wire. Sit one of your pupils on the handle of the pump and steady him. Turn the water tap on slowly and see if the water pressure will lift him.

2.277 A model hydraulic ram
Hydraulic rams are sometimes used to raise water from a low level to a higher level. They are operated by a flowing stream of water. You can make a model hydraulic ram in the following way. Obtain a plastic bottle from which the bottom has been removed. Fit the bottle with a one-hole rubber stopper carrying a short length of glass tubing. Connect this to a glass or metal T-tube which has a piece of rubber tubing on one end and a jet tube connected to it with a rubber tube as shown in the diagram. Fill the bottle with water and pinch the tube at the end. Let the water run from the end of the tube. Stop the flow suddenly by quickly pinching the tube, and note the height to which the water squirts from the jet tube. Let the water flow and stop alternately, and you have a working model of the hydraulic ram.

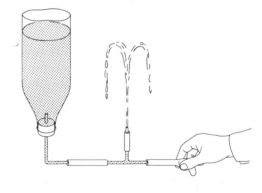

2.278 Model water wheels
Model water wheels can be made from old typewriter spools or sticking plaster reels. The blades are made from pieces of tin cut as shown in

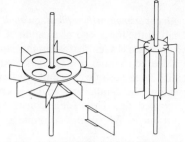

the figure above and soldered inside the spool. A meat skewer or a knitting needle is used as an axle. A stream of water from a tap, or guided from a tank along a piece of rainwater spouting is a suitable source of power. A cotton reel or cork makes another kind of wheel. Cut slots down the sides perpendicular to the ends as shown above. Slide rectangular pieces of wood or tin into these slots to act as paddles.

Buoyancy

2.279 *The buoyancy of water*
Find a metal can like a coffee can or a cigarette tin which has a tightly fitting cover. With the cover on, push the can into a pail of water, cover end down, and quickly let go of it. Repeat this having

the can in different positions. What do you observe? Can you observe the upthrust on the can? Put a little water in the can and repeat the experiment. Keep adding water a little at a time and repeating until the can no longer floats. Fill the can with water and put the cover on. Put a double loop of string around the side of the can and then attach a large rubber band to the other end of the cord. Lift the can by holding the rubber band and observe how much the band stretches. Now lower the can in a pail of water and observe the stretch in the rubber band. How do you account for the difference?

2.280 *A Cartesian diver*
Find a tall glass jar with a fairly wide mouth. Wrap a few turns of copper wire about the narrow part of the rubber bulb from a medicine dropper. Fill the jar brimful with water. Put some water in the bulb and float it in the jar of water. Add enough water to bring the bulb nearly to the point of sinking. Considerable adjustment will be required for this, and at this point almost all the rubber will be under water. Remove air from the bulb a bubble at a time by pinching the bulb. When you have adjusted the diver, put a solid stopper in the jar or tie a piece of rubber cut from an old inner tube over it. By pressing on the stopper or rubber, the diver will sink. When the pressure is released, it will rise to the surface. If you make the floater from a small glass test tube or a medicine vial you can explain the action of the diver by observing the level of water inside the float when it sinks and when it floats.

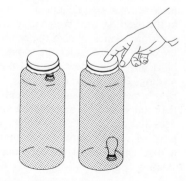

2.281 *Sinking bodies*

You will require an overflow can, a stone that will go inside the overflow can and a catch bucket made from an old tin. Fill the overflow can with water to the level of the spout. Attach a string to the stone and weigh it with a spring balance. Weigh the catch bucket and place it underneath the spout of the overflow can so that it will catch

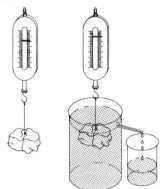

the water displaced when the stone is put in the water (see figure). Immerse the stone in the water and record its weight. Does it weigh the same as in the air? Find the weight of the displaced water by subtracting the weight of the bucket from the weight of the bucket and water. How does the apparent loss of weight of the stone compare with the weight of the displaced water? Try this experiment with other sinking bodies.

2.282 *Floating bodies*

Fill an overflow can with water and let it run out until the surface is level with the spout. Select a piece of wood that floats half or more submerged in the overflow can. Weigh the piece of wood with a spring balance. Weigh the catch bucket. Place the catch bucket under the spout. Put the wood block in the overflow can and note the balance reading. Find the weight of the displaced water by subtracting the weight of the catch bucket from the total weight of catch bucket and water. How does the apparent loss of weight of the floating piece of wood compare with that of the water it displaces? Repeat the experiment with other floating bodies.

2.283 *An experiment with a floating candle*

Put a nail in the lower end of a candle. Choose a nail whose mass is such that when the candle and nail are floated in water in a tall glass, the candle floats with its top a little above the surface of the water. Light the candle and watch it until it is nearly burned up. The candle constantly loses mass as it burns. Why does it continue to float?

2.284 *A floating experiment with different kinds of wood*

Obtain a cork, and pieces of different woods such as maple, mahogany and ebony (see figure). Place them in a pan of water and notice the way in which each piece of wood floats. Can you explain this?

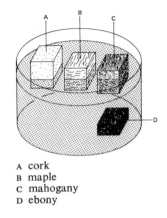

A   cork
B   maple
C   mahogany
D   ebony

2.285 *An experiment with a floating egg*

Place an egg in a glass of fresh water and observe it. Now add salt to the water and see if the egg will float (see figure). Can you explain this? How does this relate to the fact that ships ride higher in ocean water than they do in fresh water?

2.286 *Drinking straw hydrometer*
Obtain a drinking straw or a stout natural straw about 20 cm long. If it is not 'waterproof' dip it in melted candle wax and allow it to dry. Seal one end with wax, and introduce some lead shot or fine sand until it floats in water in a vertical position. Then drop in melted wax to keep the shots or sand in place. Place a thin rubber band or a piece of black cotton round the stem so that it can be slid up and down as a marker. Put a mark on the straw at the water level. Then take the straw out of the water, and measure the length of the straw from the bottom to the water level mark. Let it be x cm. Now let us assume that water has a relative density of unity and that the straw has a uniform area of cross-section. Thus we may put a set of markings on the straw for measuring relative densities of different liquids with ranges, say, from 0.6 to 1.2 by using the formula:

length of straw from the bottom to the mark
$$= \frac{x}{\text{relative density of liquid}}$$

2.287 *Floating in different liquids*
Obtain a tall, slender glass jar, test-tube or bottle, and the following liquids: mercury, carbon tetrachloride, water and kerosene. You will also need a small iron or steel ball such as a ball bearing, or iron nut or bolt; a small piece of ebony or some other wood that sinks in water; a piece of paraffin wax; and a piece of cork. First pour some mercury into the glass jar, then some carbon tetrachloride, some water and some kerosene. Drop the four solid substances in and you will observe that the iron sinks in the three top liquids but floats on the mercury. The ebony sinks in two liquids but floats on the carbon tetrachloride. The paraffin sinks in the kerosene but floats on the water and the cork floats on the kerosene.

2.288 *How a submarine is raised and lowered*
Place pieces of iron or stones in the bottom of a small wide-mouth bottle and pour a little melted paraffin wax on them to fasten them down so that the bottle will float in an upright position. Insert a two-hole stopper. In one hole place a U-shaped length of glass tubing which extends to the bottom of the bottle. In the other hole put a short length of glass tube; join a rubber tube to this. Put the bottle in a vessel of water. Withdraw some air by sucking on the rubber tube and water will siphon into the bottle until the bottle sinks. The bottle may be made to rise by blowing out part of the water.

Because the use of compressed air to empty the tanks is not practical while the submarine is submerged, submarine engineers adjust the buoyancy of a submarine to that of the water and then use elevators to dive or climb. To remain at the surface, they 'blow' the tanks empty with surface air after rising.

The device also illustrates the principle of the tanks or pontoons used to lift sunken ships. Fasten a weight to the bottle, sink both in water and lift the weight by blowing air into the bottle.

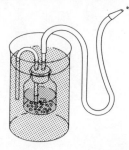

2.289 *Sinking and floating*
Shape a piece of lead, tin or aluminium foil into the form of a little boat and float it on water in a pan; now squeeze the metal foil boat into a small ball and try to float it on the water. What do you observe? What is your best explanation for this?

A   cork
B   paraffin wax
C   ebony
D   iron or steel
E   kerosene
F   water
G   carbon tetrachloride
H   mercury

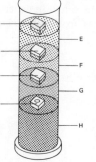

Surface tension

2.290 *The effect of soap on surface tension*

Select a large plate and rinse it until you are sure that it is very clean. Fill the plate with cold water and let it stand for a time on the table until the water is still. Sprinkle some talcum powder lightly over the surface of the water. Wet a piece of soap in water and touch it to the water near the edge of the plate. The talcum powder will be drawn to the opposite side of the plate at once. The soap reduced the surface tension at one point and the increased surface tension on the other side contracts the surface and pulls the talcum with it. Try a similar experiment but substitute flowers of sulphur for the powder and synthetic liquid detergent instead of the soap. If a transparent dish is used it can be placed on an overhead projector and the results displayed on a screen.

2.291 *Floating a needle on water*

Obtain a steel needle and dry it thoroughly. Place it on the tines of a dinner fork and gently break the surface of some water in a dish with the fork. If you are careful, the needle will float as you take the fork away. Look at the water surface closely. Can you see how the surface film seems to bend under the weight of the needle?

2.292 *Floating a razor blade*

Obtain a razor blade of the double edge type. Try floating it on the surface of water. Again observe the surface and see if the surface film dips under the razor blade.

2.293 *Lifting the water surface*

Bend the pointed end of a pin or use a piece of fine wire to make a hook. File the point of the hook until it is very sharp. Put your eye on a level with the surface of the water in a drinking glass. Put the hook under the surface of the water and gently raise the point to the surface. If you are careful, the point will not penetrate the surface film but will lift it slightly upwards.

2.294 *Holding water in a sieve*

Pour some oil over the wire mesh of a kitchen sieve and shake out the excess so that the holes are open. Obtain a pitcher of water and carefully pour it into the sieve by letting it run down the side of the sieve. When the sieve is about half full, hold it over a sink or pail and observe the bottom. You will see water pushing through the openings but the surface tension keeps it from running through. Touch the bottom of the sieve with your finger and the water should run through.

2.295 *Heaping water up in a glass*

Place a drinking glass in a shallow pan or on a saucer. Rub the top edge of the glass with a dry cloth. Pour water into the glass until it is full to the brim. You will observe that you can fill the glass several millimetres above the top. Now drop coins or thin metal washers into the water edgewise. By dropping these in see how far you can heap the water up before it runs over.

2.296 *Pinching water*

Obtain a used tin can and make five holes in it with a nail. The holes should be very near the bottom of the can and about 5 mm apart. Now fill the can with water and observe that the water comes from the can in five streams. Pinch the

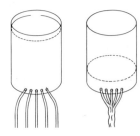

jets of water together with your thumb and forefinger and you can make one stream from five. If you brush your hand across the holes in the can, the water will again flow in five separate streams (see sketch).

2.297 *Driving a boat by surface tension*
Obtain some gum camphor or moth balls. Cut two or three boats from stiff paper, each about 2.5 cm in length. Cut a notch in the stern large enough to hold a small lump of gum camphor in contact with the water without letting it fall out. Float your boats in a large pan of water. You can make an interesting variation by placing the notch in the stern on the right or on the left (see figure).

2.298 *Blowing soap bubbles*
Soap films and bubbles serve very well for observations of surface tension. You can make a good soap bubble solution by placing three level tablespoonfuls of soap powder or soap flakes into four cups of hot water. Let the solution stand for three days before using. Try blowing bubbles with a bubble blower, a soda straw, a clay pipe and an old tin horn about 4 cm in diameter. Another good bubble blower can be made by slitting the end of a soda straw into four parts extending about 1 cm from the end. Bend these pieces outwards. A razor blade works well for slitting the end.

2.299 *A soap bubble support*
Put a round dowel rod about 15 cm long into a wooden base. Wind copper or iron wire around the dowel rod and make a loop about 10 cm in diameter (see figure). Dip the loop in soap solution. Blow a large soap bubble and place it in the loop. Now wet a soda straw in the soap solution and carefully put it through the large bubble. Try to blow a smaller bubble inside the large one. This will take a little practice.

2.300 *Experiments with soap films*
Make the forms shown in the figure from wire. Dip the various forms in the thick soap solution and observe the films. Dip the wire form with

the slider in the soap solution. Pull out the slide slightly and watch the film stretch. Release the slider, and it will be pulled back by the contraction of the film.

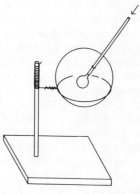

2.297   Driving a boat by surface tension

2.299   A soap bubble support

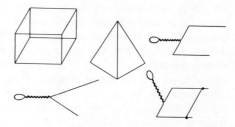

2.300   Experiments with soap films

Atmospheric pressure

2.301 *Pressure experiments with syringes*
Plastic syringes of the 100 cm³ size can be used
for a number of investigations with air pressure.
When the tip is sealed, the syringe can be used
to compress the air or to produce a partial
vacuum. Attaching a small piece of plastic
tubing makes it easy to seal off the tip with
pinch clamps or wooden plugs. A syringe can
also be sealed by pushing the tip into a wooden
or plastic block that has been drilled to the
proper size. With such a base as a platform the
syringe can then be used in a vertical position
with applications such as serving as a balance
for measuring weight by air compression. Filled
with a small amount of air and hung inverted
from screw-eyes, it can serve as a 'spring-type'
balance. Compressing moist air within a syringe
will cause water condensation and form 'rain'.
Attaching a piece of plastic tubing 20 or 30 cm
long furnishes a simple syringe pump. Using
various lengths and putting water in the tube
can result in an air thermometer, or thermo-
barometer, or if 11 or 12 metres of tubing are
used, a water barometer. Coupling two syringes
with a piece of tubing allows demonstration
of pressure changes within closed systems (see
figure). Quantification of all these experiments
is easily possible since the syringes are already
graduated. (See also experiments 2.196 and
2.309.)

2.302 *Finding air*
Plunge a narrow-necked bottle mouth down
into a jar of water. Slowly tip the mouth of the
bottle toward the surface of the water. What
do you observe? Was the bottle empty? Place
a lump of soil in a container of water and observe.
Did you see anything that might indicate the
presence of air in the soil? Find a brick and
place it in a container of water. Is there any
evidence that air was inside the brick? Fill a
glass with water and observe it closely. Let the

glass stand in a warm place for several hours.
Observe again. What difference do you see?
Is there any evidence that water contains air?

2.303 *Air takes up space*
A. Obtain a bottle and a funnel. Place the funnel
in the neck of the bottle. Fill the space around
the funnel with modelling clay. Be sure to pack
the moist clay tightly in the neck of the bottle.
Pour water slowly into the funnel (see figure).
What do you observe? What does this show
about air? Repeat the experiment and pour
in water until it comes nearly to the top of
the funnel. Carefully punch a hole through the
modelling clay into the inside of the bottle with a
nail. What did you observe? Why did it happen?

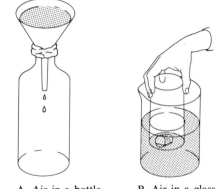

A. Air in a bottle   B. Air in a glass

B. Pour water into a large glass jar until the
jar is half full. Float a cork on the water and
lower a drinking glass, mouth downward, over
the cork. What do you observe? Wedge a
piece of paper tightly into the bottom of the
glass and repeat the experiment. Does the paper
get wet?

C. Obtain an aquarium or a big water bowl
and fill it nearly full of water. Lower a drinking
glass, mouth downward, into the aquarium.
With your other hand lower another glass into
the aquarium. Let this second glass fill with
water by tilting its mouth upwards. Now hold

this glass above the first one mouth downwards. Carefully tilt the first glass to let the air escape slowly (see sketch). Fill the second glass with air from the first glass. What does this show about air?

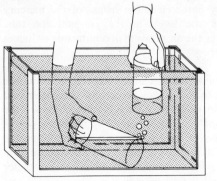

C.   Transferring air under water

2.304 *Air has mass*

Place a flat stick about a metre long so that nearly half of it hangs over the edge of a table or desk. Lay a full sheet of newspaper over the end of the stick on the table and smooth it down carefully. Give the other end of the stick a sharp blow with your hand or a wooden mallet. The stick will break over the edge of the table. The stick breaks because the inner end has been held down by air pressure on the large sheet of paper. Stand to one side when hitting the stick. (See also experiment 4.116.)

2.305 *Air exerts pressure*

A. Fill a drinking glass to the brim with water. Place a piece of cardboard over it. Hold the cardboard against the glass and turn the glass upside down. Take away the hand holding the cardboard (see diagram). Place the inverted glass on a smooth table top and carefully slide it off the cardboard on to the table top. Move the glass slowly over the table top. Can you suggest a way to empty the glass without spilling the water on the table top? What does this experiment show about air? (See also experiment 4.117.)

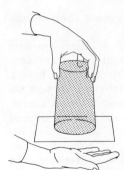

Air supports water in a glass tube

Air supports water in a glass

B. Hold a finger over the end of a piece of straight glass tube or soda straw and lower it into a jar of coloured water. Remove the finger and observe what happens. Replace the finger on the top of the tube and then lift the tube from the jar (see figure). What happens? Why? What does this show about air?

C. Make a hole with a nail near the bottom of a tin can. Fill the can with water. Hold the palm of the hand tightly over the top and water will stop running from the hole. Remove the hand and water runs from the hole (see sketch). What does this show?

D. Wet the bottom of a plumber's force cup and press it against some flat surface such as the top of a stool. Try to lift the stool with the stopper. Why is this possible? Wet the rims of two plumber's force cups. Press the rubber cups tightly together and then try to separate them (see figure). Why is it so difficult to pull them apart? This experiment is similar to the classic Madgeburg Hemispheres experiment.

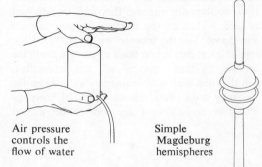

Air pressure controls the flow of water

Simple Magdeburg hemispheres

2.306 *Using air pressure to put a straw through a potato*

Place the index finger over one end of a soda straw and hold a potato in the other hand. Rapidly jab the straw through the potato, being careful to hit the potato squarely with the straw. By sealing the top end of the soda straw with a finger, the air in the straw is trapped when the other end of the straw strikes the potato. This compressed air gives the straw strength enough to prevent its bending. The result is sudden and surprising. The fragile straw will easily go through the potato (see sketch).

2.307 *A simple mercury barometer* (before carrying out this experiment refer to the section on handling mercury in Chapter One)

A. Seal one end of a glass tube about 80 cm in length by rotating it in a gas flame (diagram A opposite). The tube should be held as nearly vertical as possible. Attach a small funnel or thistle tube to the open end of your barometer tube with a short length of rubber tube. Pour mercury into the tube slowly. *Caution.* Mercury vapour is dangerous and should not be inhaled. If air bubbles are trapped, they may be removed by gently shaking the mercury in the tube up and down. Fill the tube to within 1 cm of the top. The last part is best filled by using a medicine dropper so that mercury will not be spilled. Fill the tube until a little mercury extends above the tube level. Pour about 2 cm of mercury into a bottle or dish. Place your rubber gloved

finger over the end of the tube and place the tube open end down in the jar of mercury. Remove the finger from the tube when it is under the surface of the mercury. When this tube is properly supported it will serve as a mercury barometer. The height of the mercury between the levels in the jar and the tube measures the air pressure in centimetres or inches of mercury.

B. An ink bottle can be used as a container for the mercury in a permanent barometer and will help keep the surface clean and reduce the escape of mercury vapour. The following procedure may be used to set it up. Before filling the tube with mercury as described in A above, find a cork with two holes, one to fit the barometer tube, and the other to fit a short glass tube (see diagram B). Slide the cork on to the tube until it is about 15 cm from the open end and insert the short tube in the other hole. Now stick a rubber cycle patch on to the bottom of the

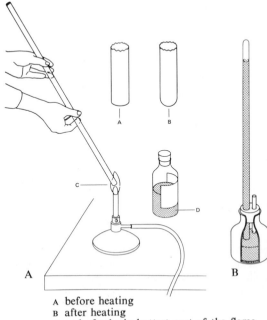

A before heating
B after heating
C end of tube in hottest part of the flame
D mercury

bottle just opposite the mouth. Fill the barometer tube as described, and place the bottle neck downwards over the open end, pressing the patch hard on to the top of the tube. Keeping the tube in contact with the patch, turn both over and stand the bottle on the bench. Still pressing on the tube, pour some mercury into the bottle. Now raise the tube a little to allow mercury to run from the tube, and push the cork into the neck of the bottle. If desired, the barometer may now be supported in a bracket with a metre scale attached to it and hung on a wall. The top of the barometer tube should then be supported, and the ink bottle can be made to fit tightly in a tin fastened to the bracket.

2.308 *An aneroid barometer*

A corrugated rubber tube from an automobile, or a cycle handle grip, can be used to make a model aneroid barometer. No great accuracy is to be expected because of the many possible errors. Two good corks or pieces of non-porous wood are needed to close the ends of the tube, which serves as a vacuum box. They are fitted when the rubber is compressed and they should be made airtight with wax and by tying round the outside of the rubber. A weight hung from the lower cork will partially counteract the result of atmospheric pressure and extend the bellows. If a pointer is mounted as shown in the figure a scale can be attached to the apparatus to give readings of the variations in atmospheric pressure.

2.309 *Measuring atmospheric pressure with a bicycle pump*

A bicycle pump with the washer reversed as illustrated below can be used to measure atmospheric pressure. The piston can be made airtight by adding a little thick oil to the barrel. The area of cross section of the pump barrel can be calculated or measured with squared paper. The pressure of the air can then be calculated in kg/cm². The weight supported by the upthrust of atmospheric pressure is found by hanging various loads from a hook, screwed into a wooden plug fitted into the pump handle. (See also experiments 2.196 and 2.301.)

2.310 *Measuring atmospheric pressure with a rubber suction cup*

The force required to pull the sucker away from a smooth surface can be found by using a spring balance. The area on which the atmospheric pressure is acting can be measured by pressing the sucker on a piece of squared paper.

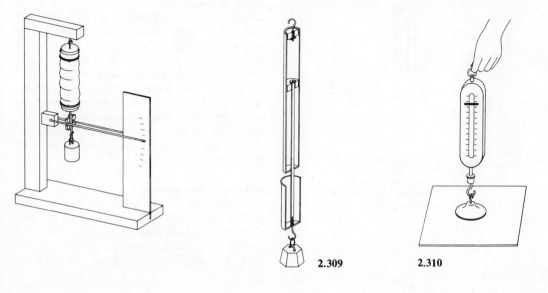

2.309   2.310

Preferably use a sucker which has a hook attached. If one of this sort is not available, tie some
copper wire firmly round the neck of the sucker
to be used and form a loop. If the laboratory
bench is not smooth enough use a piece of
plate glass, holding this down with one hand
whilst pulling on the spring balance with the
other. Make several trials and, if possible, use
suckers of different sizes (see figure, page 141).

2.311 *A simple syringe lift pump*
Assemble a simple syringe using glass or metal
tubing (iron pipe or conduit tubing is suitable),
two corks and a piece of metal rod. The cork
which serves as the piston is made to fit tight
by wrapping a string round it. The other cork,
with a piece of glass, bamboo or strong tubing,
acts as an intake. Burn two holes through the
piston with a hot wire and fit a thin piece of
leather above them to act as a valve which closes
on the upstroke and yet allows liquid to pass
through on the downstroke (see figure).

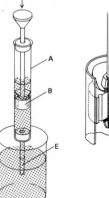
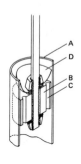

A glass or metal casing
B cork piston
C holes in piston
D leather or rubber valve
E intake tubing

2.312 *A test-tube force pump*
To make this apparatus, heat the bottom of
a test-tube with a small flame and blow a hole.
Now blow a hole in a larger test-tube and fit
both with ball bearings or small marbles to act
as valves. Wrap string around the inner test
tube so that it fits tightly in the outer one but
can still slide up and down. Fit a cork and a tube
securely in the inner tube as shown in the diagram. It will now serve as the piston of a force
pump (see sketch).

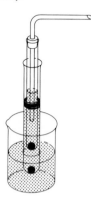

2.313 *A simple siphon*
Obtain two tall glass bottles and fill each about
half full of water. Connect two 30 cm lengths
of glass tube with a 30 cm length of rubber or
plastic tubing. Fill the tube with water and
pinch it. Put a glass tube in each bottle of water.
Siphon the water back and forth by varying the
height of the bottles. The experiment is more
interesting if the water is coloured with a little
ink. Place the two bottles on a table. Does the
siphon flow? Can you explain how air pressure
helps the siphon work?

2.314 *A siphon fountain*
Fit a glass jar (or a flask made from a used
electric bulb) with a two-hole rubber stopper.
Through one hole place a jet tube which will
extend to about half way to the top of the
flask and about 2 cm outside the stopper.

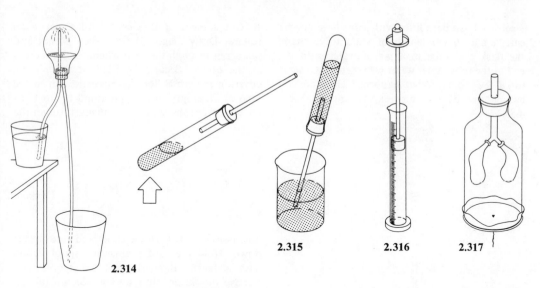

2.314

2.315   2.316   2.317

Through the other hole push a short length of glass tube so that it is just flush with the bottom of the stopper. Connect a 20 cm length of rubber tube to the jet tube. Connect a 1 m length to the other glass tube. Place some water in the flask and insert the stopper. Put the short rubber tube in a container of water on a table, let the longer rubber tube go into a pail on the floor and then invert the siphon (see figure). The fountain can be seen better if the water in the jar on the table is coloured with a little ink. You can make a double siphon fountain by making another flask unit similar to the first one and connecting them together.

2.315 *Lifting water with air pressure*
Fit a test-tube with a one-hole cork and glass. tube. Drive out the air by boiling a little water in it. Invert it with the open end under the surface of a jar of water. Atmospheric pressure will drive water upwards until it almost completely fills the test tube (see figure).

2.316 *The relation between volume and pressure of air*
Obtain a rubber bung which just fits inside a narrow glass jar or measuring cylinder. Attach

it to the lower end of a wooden rod. Fit a tin lid to the upper end of the rod to act as a scale pan. Lubricate the piston so formed with a little vaseline or heavy engine oil. Use the piston to trap air in the jar; put different weights on the pan and measure the volume of air inside the glass cylinder for each weight (see diagram). Note that the volume is in inverse proportion to the pressure.

2.317 *Working model of the lungs*
Cut the bottom off a large plastic or glass bottle. Fit a cork to the neck with a Y tube in it. On each of the lower limbs of the Y tube tie a rubber balloon or some small bladder. Tie a sheet of brown paper or sheet rubber round the bottom of the jar, with a piece of string knotted through a hole and sealed with wax. Pulling this string lowers the diaphragm and air enters the neck of the Y piece causing the balloons to dilate. Pressing the diaphragm upwards has the opposite effect (see diagram).

2.318 *Oxidation and air pressure*
Wash a small wad of steel wool in petrol or benzine to remove any grease. Squeeze it out and then fluff it. As soon as it is dry, place the

steel wool in a flask fitted with a one-hole stopper carrying a 40 cm length of glass tube. Stand the flask and tube in a jar of water with the end of the tube under water (see sketch). Observe for a few hours. What happens? How do you account for it? (See also experiments 2.40 and 4.58.)

2.319 *Air stream experiments*
A. Place a ping pong ball inside a funnel. Blow hard through the stem of the funnel and see if you can blow the ball out of the funnel. Invert the funnel and hold the ping pong ball in the end. Blow hard through the stem and see what happens as you remove your hand holding the ball. Place the ball on a table. Cover it with the funnel. Blow through the stem and see if you pick the ball up from the table. How do you explain your observations? (See diagram A below.)

B. Cut a piece of thin cardboard about 7 cm square. Draw diagonals from each corner and put a pin through the card where the lines cross at the centre. Secure the head of the pin by covering it with Sellotape. Place the pin in the hole of an empty thread spool and try to blow the card from the spool by blowing through the

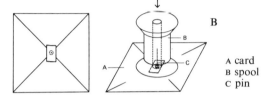

A card
B spool
C pin

hole (see figure). Turn the spool and card upside down. Hold the card against the spool lightly with a finger. Blow through the spool, then remove the finger. How do you account for the result?

C. Attach a funnel to a source of compressed air such as a vacuum cleaner. Blow up a balloon and place a piece of copper wire around the neck for a weight. Turn on the compressed air and balance the balloon in the air stream. Try also to balance a ping pong ball between the balloon and the funnel (see diagram C below).

D. Obtain two glass tubes or two transparent soda straws. Place one tube in a half glass of coloured water. Place the second tube at a right angle with the first one so that the ends of the two tubes are close together. Blow through the horizontal tube and observe the water level in the second tube (see diagram D). How do you account for the result?

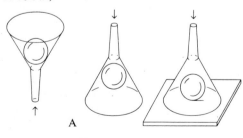

A

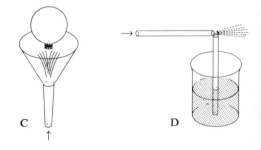

C   D

Chapter Three

Biological sciences

Introduction

In any region of the earth, teachers and students are surrounded by an inexhaustible supply of living things worthy of scientific study. Unfortunately, few teachers exploit the opportunities provided in their local situation. There are many reasons for this failure to use the environment as a source of materials for science classes. Perhaps this occurs because most teachers have not discovered an organizational scheme that enables them to relate daily observations of living things to a personal generalization regarding life. One example of such a conceptual scheme is presented here. Specific activities are then organized in accordance with the unifying generalization. This is provided with the hope that it will not only enable teachers to organize observations and encourage them to observe constantly, but will also make them feel at ease in their work with living things.

Levels of organization

The major generalization employed here is described by biologists as the 'levels of organization' approach to the study of living things. Inherent in this way of thinking is the notion that life can best be understood by organizing living things, groups of living things, and parts of living things into a natural order or hierarchy. In using this scheme in science teaching emphasis is placed upon the centre of the hierarchy, since this is the position of whole living things (organisms) which constitute most of the experiences that people have with life.

Explanation of levels

A complex subject becomes much less confusing when one begins to imagine a general pattern that permits the arrangement of a great many isolated observations into a few logical categories. Teachers of science cannot avoid being overwhelmed by the infinite variety of living things and their activities. However, science teachers must advance beyond the stage of merely appreciating nature's complexity. They must arrange their observations on the basis of some fundamental scheme that leads to increased understanding of life. One such scheme is called the 'levels of organization' concept. The scheme or model contains many parts which are described in the following sections.

3.1 *Organisms*

Organisms are individual life forms, many of which are commonly experienced. An individual dog, tree, fish, earthworm, mushroom, or yeast cell can be called an organism. Organisms vary considerably in size. A whale may be 10 million times larger than a single bacterium. Two important observations regarding organisms are: (a) they are non-uniform in internal structure; and (b) they exist with others of their kind. These two

Conceptual scheme

Higher levels	Biosphere Biomes Communities Populations	Groups of organisms
	Organisms	
Lower levels	Organ systems Organs Tissues Cells	Organisms or parts of organisms
	Organelles Macromolecules Molecules Particles	Parts of organisms

observations obviously suggest an examination of the parts of organisms (lower levels of organization) and groups of organisms (higher levels of organization).

3.2 *Higher levels*
A. *Populations*. A group of organisms comprising all of a particular kind is called a population. Usually a sub-population is described in terms of some space which it occupies. For example, one may refer to the snail population in a classroom aquarium, or the population of that kind of snail in a pond. However, if no space is mentioned it is assumed that the population consists of all snails of that type in the world.

B. *Communities*. Populations do not exist in isolation. They are commonly found in an environment which they share with other populations. All the populations within a defined space constitute a community. A lake community consists of all the plant and animal populations found in and on the body of water. The populations found in a school grounds would be a community.

C. *Biome*. Certain large areas of the earth contain communities which are similar. This collection of similar communities is called a biome. A biome may occupy a large portion of one of the continents. For example, a grasslands biome occupies much of the central portion of North America. Climate and topography are fairly uniform across a biome.

D. *Biosphere*. Life on the earth is normally found within a few metres of the surface. This hollow spherical space is called the biosphere. It contains all life on the planet. The question of whether or not there are other biospheres is not yet answered. This indicates that upward extension of levels of organization is unknown. As space exploration reveals additional information, a next higher level may be discovered.

3.3 *Lower levels*
A. *Organ systems*. Some animal organisms contain systems of organs that perform vital functions. The circulatory system consisting of the heart and blood vessels is an example of such a system.

B. *Organs*. Not all organisms contain organ systems. Plants and many animals do not appear to have distinct systems. Most, however, contain basic structures called organs which in turn are composed of tissues. A heart is an organ as is a leaf, a lung or a root.

C. *Tissues*. A tissue is a group of similar cells performing a single function. Muscle tissues, for example, are composed of cells that are capable of contracting and producing the 'pull' of the muscle. Some organisms are composed of tissues, but apparently do not have organs.

D. *Cells*. Tissues are made of individual units called cells. Just as the dollar, pound or mark have become the basic structures of national money systems, the cell is the fundamental unit in most organisms. Cells vary considerably in size from the largest, an ostrich egg, to one of the smallest micro-organisms. Cells also vary in their function. Despite these differences, there are common features that have resulted in cells receiving considerable attention as biologists attempt to understand life. There are organisms that are composed of a single cell; they are called unicellular organisms.

E. *Organelles*. The invention of the microscope led to the discovery of cells, and improved instruments revealed that cells contain parts commonly called organelles. Prominent organelles are easily seen with a light microscope. It was the invention of the electron microscope, however, which enabled biologists to develop a comprehensive understanding of cell parts.

F. *Macromolecules*. The electron microscope and other advanced technological instruments, such as X-ray diffraction devices, enabled biologists to gather information about the structure of cellular organelles. It was found that organelles are composed of large molecules (macromolecules) such as proteins, lipids (fats and oils) and nucleic acids (DNA and RNA).

G. *Molecules*. Macromolecules are long chains of linked individual molecules. A molecule is the smallest possible piece of a substance which retains the properties of the substance. Molecules are composed of atoms joined or bonded together. An atom is the smallest part of an element.

H. *Particles*. Atoms are composed of fundamental particles such as protons, neutrons, electrons, etc. This is the present limit of understanding of organization at the lower level. It is important to observe that at both the upper level of organization (the biosphere) and the lower level there is uncertainty regarding the possibility of another level not yet discovered. It is important for a science teacher to be aware of the uncertainty of the extent of man's knowledge of 'life'. Students will study life most frequently at the central levels of organization, near the level occupied by organisms. They need to understand, however, that scientific knowledge is subject to revision as experiments produce new information.

Studying organisms

Why children should study living organisms

Modern biology emphasizes the study of living organisms rather than sacrificed or preserved specimens. Thus a reasonable beginning for children in the study of living things is to start with those things that are alive. Whenever possible this is best accomplished in the organism's natural environment. A logical introduction is the study of the behaviour of an organism.

Bird behaviour

Most children have had casual experience of observing birds. However, nearly all such experiences occur without adequate preparation for careful observation and measurement. The chance for a significant discovery is greater when a child is prepared and motivated to learn something about the birds' behaviour. Some guidelines on preparation in the classroom are given below.

3.4 *Beak types and functions*

A study of beak types is followed by observation of the feeding behaviour of a bird with a particular beak type. By combining all the observations of students in a class, many beak uses may be revealed. For beak types not observed, encourage the pupils to infer the respective functions. (See illustration.)

3.5 *Feet types and functions*

Observations of birds reveal many uses of feet. Some of these uses are for wading, walking, swimming, perching, hunting, and carrying objects. A walk along a lake or stream may produce not only visual observations but also an opportunity to prepare permanent casts of imprints in soft soil or mud. Take along some cardboard (lightweight posterboard), clasps, a small sack of plaster of Paris (patching plaster), a pan and mixing spoon. Form a cylinder from the cardboard, securing it with a paper clasp or rubber band. Place it around a track and pour a mixture of plaster into the top of the cylinder. When the plaster hardens, a raised or negative print of the track is formed. This may be used to prepare a positive print if desired (see below).

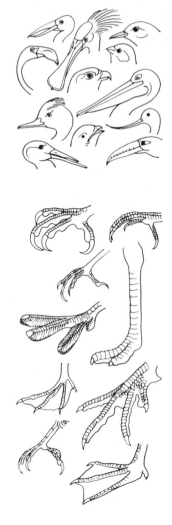

Beak types

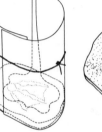

Feet types

Making casts of imprints

A collection of casts may be organized into categories based upon use or function. Water fowl, such as ducks or pelicans, display webbed feet useful in wading and swimming.

Large talons may indicate that the bird uses its feet in hunting. Hawks and owls are examples.

Many birds use their feet in perching or grasping, while others use them primarily for walking. Woodpeckers are an example of the former and quail exemplify the latter.

3.6 *Nesting behaviour*
When nests can be observed, a variety of activities may be pursued. Have children observe adult nesting behaviour. After hatching, observe the feeding and care of young. Nest building can be followed by observing bird habits and also noting materials used. Abandoned nests reveal details of construction and quite often small organisms that live in the nesting materials. Defence of its territory by a bird is an important behaviour in limiting the bird population in a given area. Have children look for aggressive bird behaviour in nesting and feeding areas.

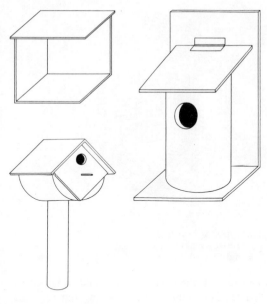

3.7 Nesting houses

3.7 *Using houses to attract birds*
Birds may be attracted by houses which will facilitate their nest-building behaviour. The house should be constructed with the following suggestions in mind:
1. The internal size must be suitable for the nest of the bird you wish to attract.
2. The entry hole must be of appropriate size.
3. The interior should be left unpainted.
4. The house must be placed where the bird will use it, and located at the proper height above the ground.

Small birds are lured by small houses with tiny openings. A wren house should be approximately 10 cm by 10 cm by 12 cm, with an opening hole 2 to 2.5 cm in diameter (see lower sketch).

Some birds such as the robin will nest in an open structure (see top sketch).

Some birds such as screech owls require a house that resembles a tree trunk. It should have an opening 10 cm across to accommodate a large bird (see sketch).

3.8 *Using feeders to attract birds*
Bird study can continue indefinitely if pupils construct a bird feeder for the schoolroom windowsill or school yard. An offering of mixed seeds and another of suet will attract a variety of bird types at all seasons of the year. Suet should be tied securely to a branch or support or placed in a cage made of metal netting. Bird feeders attract not only birds but also small mammals such as mice and squirrels. These can also be observed for indications of food preferences, times of feeding activity, and other behaviours. A cubical suet cage (see sketch overleaf) can be made from netting and nailed to a tree or post. When cutting, leave wire lengths protruding. These can be bent over adjoining squares to hold the sides together. Leave the front panel free at the top so that it may be opened to replace the suet. It can be fastened when closed with wire loops.

Pupils can make an open bird feeder from wood or metal scrap. A roof or top keeps snow

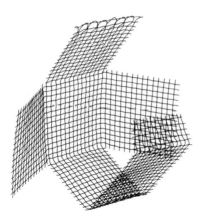

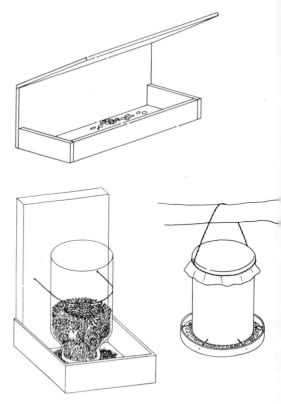

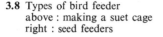

3.8 Types of bird feeder
above : making a suet cage
right : seed feeders

and rain off the seeds. Sides should be constructed to prevent birds from kicking out the seed
mixture while searching for their favourite types.
Another metal bird feeder can be made by cutting out both ends of a coffee tin, and attaching a
cake pan under it with stiff wire. Make a plastic
lid to fit over the top (see figure, right) and hang
from a branch with wire.

Aquatic organisms

An effective method of studying organisms consists of a combined effort to study them in the field
and concurrently in the classroom or laboratory.
This is especially effective with aquatic plants and
animals. Obtain or build an aquarium. It should be
made ready in advance, so that samples taken
from a visit to a pond or stream may be placed in
it upon your return.

3.9 *Jam jar aquaria*
If a large glass tank is not available, practically
any glass vessel can be used as a simple aquarium
if it is well stocked with submerged water plants
such as *Elodea* or *Myriophyllum*, to aerate the
water. A 1-kg jam jar is quite suitable for keeping
caddis larvae, pond snails, small crustacea and

plants such as *Elodea* and *Lemma minor*, and
will remain in properly balanced condition for
months if carefully stocked. It is as bad to understock as it is to overstock. The aquarium should
require no attention, but if a *Dytiscus* or other
predacious larva is kept, it should be fed regularly
on tadpoles. Three centimetres of clean sand will
provide hibernating quarters for the caddises at
the bottom of the jar, and a muslin cover will ensure that the caddis-flies do not escape unobserved. A diary should be kept to record egg laying
and other changes, as well as habits. Such an aquarium can be made the basis for a simple study of
the interrelationship between plants and animals
in pond life. For collecting pond and stream specimens, a strong net can be made from a strainer
if one is available. Its handle should be firmly
bound to a stick, the tape being threaded through

the handle repeatedly. The tape must be liberally smeared with rubber solution, if available, and then tied tightly and the knot smeared too. Avoid placing an aquarium in direct sunlight. Excessive light produces a heavy growth of algae on the glass walls, which obscures the contents of the aquarium. If algal growths occur, wipe them off with an abrasive cloth such as those used to wipe dried food from dishes.

3.10 *Large aquaria*
A glass aquarium 50 cm by 25 cm is of useful size. Old accumulator cells are suitable, but the glass is not very clear. To prepare such an aquarium take some fine silt from the bottom of a clear stream or pond and wash it carefully in running water. Cover the floor of the aquarium with it to a depth of about 2 cm. Plant a few water plants in this, weighting the roots with a stone or gravel. Then put in a layer of coarse sand or gravel and some large stones to serve as hiding places for the water insects. Fill with a slow stream of water falling on a sheet of cardboard to minimize cloudiness and allow to stand for a day or two until clear. Clean water plants should then be introduced. There is no need for elaborate aerating arrangements if plenty of water weeds are present. If tap water is used, some live food such as daphnia should be added. The animals can now be introduced, with a few snails to keep the glass clean. Very little feeding will be necessary. Fish will eat the snails' eggs and small water organisms introduced with the water plants. If worms are used as food they should be given only once a week, cut in pieces

small enough to be eaten. Any unconsumed food should be removed immediately or fungi will grow and will infect the fish. The aquarium should be covered with a glass plate or perforated lid to keep out dust. If frogs or newts are kept, a floating piece of cork must be provided for them to sit on; the glass or cover will prevent their escape.

Chick embryos

3.11 *A simple incubator*
If electricity is available in your classroom a simple incubator can be made at a very low cost. Obtain two cardboard boxes, one large and one small. Cut one end from the small box, and cut a 15 cm square window in a side of the large box. Next cut a slit in the top of the smaller box and suspend an electric lamp in it. There should be a long electric cord attached to the lamp. Place the small box inside the larger one and pack newspaper between them on all sides. Be sure the open end of the small box fits against the side of the large box in which the window was cut. Place a thermometer in the box so that you can read it through the window. A glass plate is fitted over the window.

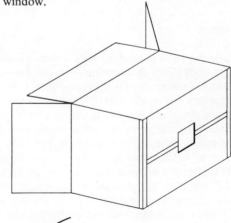

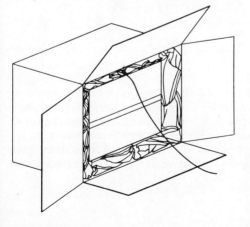

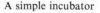
A simple incubator

3.12 *Observing the development of embryos*

It is necessary to maintain a constant temperature of 40° C in the incubator, night and day, for 21 days. By using different-sized bulbs and by changing the amount of newspaper, you will be able after a few days to regulate your incubator to this temperature. A small dish of water should be placed in the incubator.

Now secure a dozen fertile eggs. Place the eggs in the incubator and leave them. At the end of three days remove one of the eggs and crack it carefully. Put the contents into a shallow saucer. A three-day embryo will usually show the heart already beating. It may continue to beat for half an hour. Remove an egg every three days and observe the development of the embryo. Some of the eggs can be left for the full 21 days to see if any of them will hatch.

3.13 *Further observations on embryos*

Investigations can also be made into the effect on the embryo of changes in temperature. Eggs removed from the incubator simultaneously can be placed in different heat environments—the open air, refrigerator or heater, for example—and the resulting effects on the embryo noted.

Insects

The emphasis in working with insects should be the study of living specimens. Too often the capture and mounting of insects constitutes the entire experience of pupils. Certainly there is much to be learned in such an activity. Unfortunately, more valuable experiences are often omitted or neglected. Night-flying insects can be studied by setting up a light trap, consisting of a white sheet stretched out between small trees at an angle of 20°–30° from the vertical. A bright light source such as a gasoline lantern or a large dry-cell lamp is placed under the sheet so that it is brightly illuminated. During daylight hours, insect collecting requires a net.

3.14 *Collecting insects*

A useful insect air net can be made from a round stick such as a broom or mop handle, some heavy wire and mosquito netting or cheese cloth. Bend a heavy piece of wire into a circle about 38 to 45 cm in diameter, and twist the ends together to form a straight section at least 15 cm in length. Fasten this to the end of a broom or mop handle by lashing with a wire or by means of staples. Cut a piece of mosquito netting or cheese cloth to form a net about 75 cm deep. Fasten this to the circular wire frame by stitching. (See diagram below.)

It is often more informative to sweep vegetation with a net rather than to catch flying insects with an air net. A sweep net is made with canvas or muslin, instead of mesh, and heavy wire that will not bend when the net is swept through heavy grass or weeds. Have pupils 'sweep' a grassy field by working back and forth over a measured area. A count of the net contents provides a means of estimating the number of insects between the soil

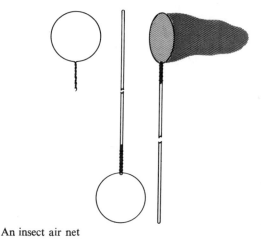

An insect air net

surface and the grass tops. A class may wish to sample the school grounds, a farm field, an abandoned field, forest floor or other natural areas to determine the relative numbers of insects.

Prior to the sweeping of vegetation, have pupils construct small cages for crickets or grasshoppers. These can be made of wood with fine wire mesh or thin wooden bars. Place grasses, water and a small dish of moist sand in each cage. Females may be observed laying eggs in the sand.

3.15 *Making a collection*
Insect collections provide a variety of studies; variation within a given type or species may be based on colour, size or other individual differences. Before starting the collections, pupils must prepare their equipment which may include nets, killing jars, stretching boards, mounting boxes, and mounting blocks.

A. *An insect killing jar.* Obtain a wide-mouth glass jar with a screw top or one which closes very tightly. Place a wad of cotton in the bottom and cover it with a round piece of cardboard or blotting paper which has several holes punched through it. When the jar is used, saturate the cotton with an insecticide. Place the piece of cardboard over the cotton and then put the insect in the jar. Close the jar tightly and leave until the insect has been killed. If moths or butterflies are being prepared, ensure that the jar opening is large enough to avoid tearing the wings.

B. *A stretching board for insects.* A stretching board is essential when insects are being prepared for mounting. You can make one easily from a cigar box. Remove the lid from the cigar box and split it lengthwise into two equal parts. Attach these to the box again, but leave a space about 1 cm wide between them. The body of the insect is placed in the slot and the wings are secured on the top by means of strips of paper held

by pins pushed into the soft wood, but not through the wings. Sometimes it is desirable to mount the top pieces at a slight angle. This can be achieved by cutting the ends of the cigar box to a slight V form before attaching the sections cut from the lid. This is shown below, left.

C. *Mounting boxes for insect collections.* Wood or cardboard cigar boxes make very useful and convenient housings for insect collections. After the insect has been removed from the stretching board, a pin is pushed through the body and pressed into the bottom of the box to hold the insect. The pins are arranged in an orderly fashion and may carry small cards giving data about the insects (see figure).

Cigar boxes may also be used to mount insects on a cotton background. The lid is removed and the inside of the box filled with layers of cotton fluff. Next the insects are arranged on the fluff and covered with glass or cellophane which is taped to the box to make a permanent mounting. This type of mounting box is especially suitable for butterflies and moths or for displays in a school museum.

D. *Mounting block guide.* A uniformly mounted collection is more attractive and makes it easy to compare specimens. Pupils should prepare a wooden mounting block which looks like three steps (see diagram). Each step has a hole drilled through its centre. The top step is used to line up all insects at the same height by impaling the insect and pressing the pin through the top hole. The other steps provide uniform levels for labels containing information about the specimens.

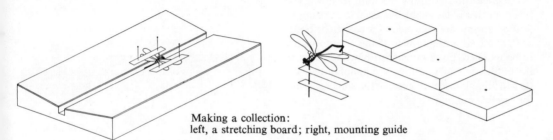

Making a collection:
left, a stretching board; right, mounting guide

E. *A simple insect cage.* Make a cubical frame from tongue depressors, ice-cream bar sticks, or dowel rods. A frame that is 15 cm on each side is convenient. Pull a nylon or silk stocking over the frame and close the open end by tying a loose knot, or by fastening with a rubber band. The open end provides access to the cage interior.

Collecting soil organisms

The collection of soil organisms provides a multitude of possibilities. The organisms may be added to the classroom terrarium. More important is their potential for introducing pupils to quantitative techniques. Standard size soil samples from various locations can be compared for total quantity of organisms or quantity of various types of organisms. A standard size soil sample can be obtained from soft ground by forcing an empty tin can into the soil to obtain a core sample. Standard surface samples of leaf litter can be obtained by means of a ring, made from stiff wire, which is placed on the litter; collect all loose litter and soil within the ring. Core samples and surface samples should be placed in plastic bags for return to the classroom. Use a soil organism funnel like the one described below to remove organisms from the soil.

3.16 *Soil organism funnel*
The funnel is made of smooth, shiny material such as the sides of a large tin. The litter is placed on a wire mesh fixed in the mouth of the funnel (a household sieve is convenient for this purpose). A 25-watt bulb with reflector is mounted over the litter (see diagram opposite).
    Caution: do not allow the bulb to touch leaf litter or other inflammable material in the sample. Soil organisms are collected in a beaker or bottle at the bottom of the funnel. If the organisms are to be retained, place moistened blotting paper in the bottle. A small quantity of ethanol in the bottle will immobilize and preserve the soil organisms for careful counting.

Trapping small mammals and reptiles

Small mammals and reptiles may be captured and maintained in cages for study. An inexpensive trap is described below.

3.17 *A simple trap*
Obtain a large glass jar with a wide mouth and a screw-on cap. A one-way door can be made by cutting an opening in the lid and attaching a free-swinging metal door that opens inward only (see diagram). The door swings on a stiff wire. Animals can be transferred to cages without direct handling.
    Caution: pupils should wear heavy leather gloves when handling reptiles or mammals. Even when non-poisonous, bites from such organisms may become infected.

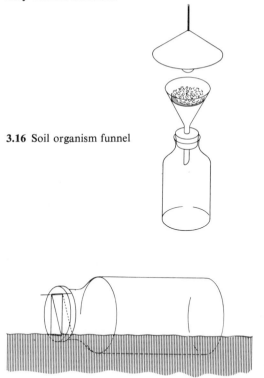

3.16 Soil organism funnel

3.17 A simple trap

Keeping animals in cages

It is frequently desirable in elementary and general science to keep animals caged in the science room for short periods of observation. To carry out such activities effectively, suitable cages and food must be provided.

3.18 *An animal cage*
Cages can be made from a variety of materials found in almost every locality. One such cage is made from a wooden box provided with a hinged lid in which a window is cut and covered with fine wire netting. Windows are also cut in three sides of the box. The side and back windows are covered with wire netting, and a glass plate is fitted in the front window. This type of cage can be improved by adding a drawer which is fitted under the front glass window and which covers the entire bottom of the cage (see diagram). This enables the cage to be cleaned without disturbing the animals to any great extent. In tropical areas very useful cages can be made using bamboo splints or other wood in place of wire netting.

3.19 *Food and water for caged animals*
In general, food and water containers should be kept above the floor of the cage. A convenient feeding trough for small animals may be made by cutting a section from the side of a tin can, bending over the sharp edges, and then attaching it to the side of the cage by means of wires, as shown in the diagram.

A watering device for animals such as mice, guinea pigs and hamsters can be made from a bottle fitted with a one-hole rubber stopper through which is passed a piece of glass or metal tubing with a narrowed opening. Invert the bottle and insert the tubing through the screen into the cage.

Regular feeding and watering of animals and regular cleaning of the cages are important, not only for the health and comfort of the animals but also for disciplined habits and a sense of responsibility in the pupils. Food and water must be changed daily and cages must be cleaned once a week.

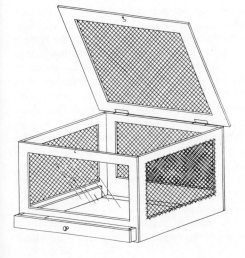

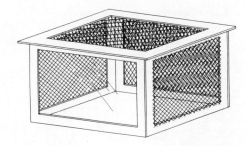

3.18 An animal cage

Flatworms

Flatworms are excellent organisms for study by pupils. They react to various stimuli, making them suitable for elementary studies of behaviour. They also possess an ability to regenerate lost parts.

3.20 *Obtaining and feeding flatworms*
Look for flatworms on the underside of submerged logs or stones in a pond or lake. The brown type (*Dugesia tigrina*) or the larger *Planaria* sp. are preferred for study. If you cannot find flatworms they can be trapped by placing a piece of raw beef liver in cheese cloth, tying it with a string and placing it in the water. Check the bait daily, and brush off the flatworms into a small jar of pond or lake water. In the classroom, the flatworms should be transferred with a large medicine dropper or pipette into opaque containers such as kitchen bowls or enamelled pans. Keep the containers covered with an opaque lid of cardboard or wood when not observing. Feed finely chopped liver, hard-boiled egg, or bits of worms once a week. After 3 hours remove excess food with a medicine dropper.

3.21 *Flatworm behaviour*
Flatworms respond to a variety of stimuli. Pupils can observe the effects of light, sound, food sources, mild electric shocks and chemicals such as a few crystals of epsom salts. A hand lens enables a pupil to observe the tube-like pharynx with which the flatworm ingests its food.

3.22 *Flatworms regenerate parts*
Flatworms *(Planaria)* can be induced to regrow parts by placing a specimen on a glass slide and cutting it with a sharp razor blade. Worms may be cut in half across the body or down the length of the body. A cut part-way down the midline of the body produces a worm with two heads, if cut from the head down, or two tails if cut from the tail end (see diagram). After cutting, return the parts to the dish and do not feed until regeneration has occurred.

Studying populations

The study of plant and animal populatsion provides pupils with experience revealing the interactions within a group of organisms of the same type or species. These experiences can occur in both the field and the classroom. The conditions observed in field-work can be compared to classroom situations where overcrowding, food shortages, oxygen limitations and other variables can be adjusted in order to study the effect on a population.

3.23 *Cultures of fruitflies*
The common fruitfly, *Drosophila*, has been widely used in genetic studies. It is easy to culture and reproduces rapidly. This makes it suitable for population studies. Fruitflies can be attracted by placing over-ripened fruit in a container. After trapping they can be transferred to small jars containing fruit chunks. Banana makes an excellent food source. Place a bit of ripe fruit in the bottom of the jar and make a paper funnel with a hole in the end to fit the mouth of the jar. Place the jar in the open, and when six or eight fruitflies have entered, remove the funnel and plug loosely with cotton wool. With this number of flies there should be both males and females. The females are larger, with a broader abdomen. The males are smaller and have a black-tipped abdomen (see sketch).

Soon eggs will be deposited, and in 2 or 3 days the larvae will hatch. A piece of paper may be placed in the jar for the larvae to crawl on when they are ready to pupate. The adult insects will come from the pupae (see diagram). By putting

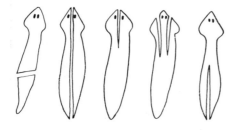

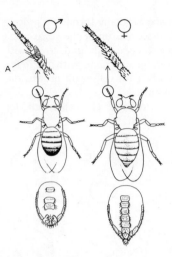

Male and female fruitflies
A sex comb

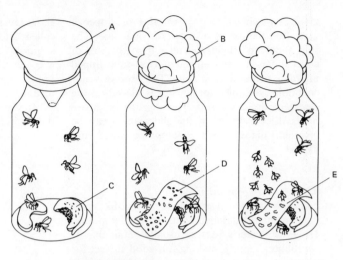

Fruitfly cultures
A paper funnel
B cotton
C eggs
D larvae
E pupae and young adults

newly hatched flies in another jar, a new generation can be started.

Cut a piece of graph paper and stand it upright in a bottle. This will enable the sampling of a large population in the bottle by counting the number of pupae on a portion of the grids. Let pupils make daily counts of the population in a bottle. When numbers become very large, estimate the total by sampling parts of the graph paper. A graph showing days along the bottom and number of flies up the left side provides a quick visual indication of the population's size. Maintain the bottle as long as flies continue to survive. Get pupils to suggest reasons for the changes in population size.

3.24 *Cultures of mealworms*
An excellent organism that may be maintained easily for long periods is *Tenebrio* sp. The larvae of this beetle are called mealworms. They can be purchased at fish-bait stores or aquarium shops.

Mealworm cultures can be very simply maintained in moist bran or oatmeal, using large jars with a screen cover which prevents the adult beetles from escaping. Adults may be fed small amounts of raw vegetables of the carrot family. Start a culture with ten mealworms. At weekly intervals, have pupils count the number of larva, pupa and adults. This activity provides a long-range source of data from a population in a limited environment. In the course of this study pupils will observe the development of beetles through distinct life stages. Within the population there will be adults, eggs, larvae and pupae.

3.25 *A hay-infusion culture*
Micro-organism populations can be cultured easily in large jars. Collect grass, leaves or other vegetation from a pond, ditch or stream. Place the material in a jar filled with water that has been boiled and allowed to cool. Take samples of the water with a medicine dropper daily for several weeks, noting the number and types of micro-organisms by viewing with a microscope. Employ the hanging drop technique for viewing. This

involves placing a ring of vaseline or stopcork grease on the centre of a microscope slide. The ring should be slightly smaller than the size of cover slip used (the open end of a test-tube dipped in grease makes a good ring). The drop of water containing the micro-organisms is placed on the centre of the cover slip. The vaseline or grease holds the cover slip to the slide which can be inverted, placed on a microscope stage and viewed.

3.26 *A yeast population*
Natural sources of yeast include the wax-like coatings on smooth-skinned fruits, especially grapes. However, baker's yeast is usually obtainable. It reproduces rapidly, making it a good subject for observing population changes under varying conditions.

A. Set up tubes of sugar, molasses or honey solutions and a water control. Place a quarter cake of crumbled yeast in each tube. Compare the results. Place a one-hole stopper with glass tubing in the sugar-yeast solution and allow the gas produced to bubble through a tube of clear lime water which will detect the presence of carbon dioxide (see diagram). (See also experiment 2.39.)

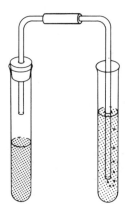

3.26A   Testing the gas produced by sugar-yeast solution

B. Yeasts reproduce asexually through a process called 'budding'. Place a drop of the sugar-yeast solution on a microscope slide and cover with a cover slip. Examine the slide with the high-power objective. Look for cells with protrusions or buds (see diagram).

If you can see nuclei in the cells, look for their presence in the buds.

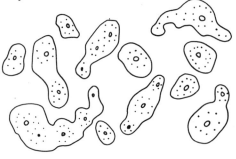

3.26B   Budding yeast cells

3.27 *Sampling yeast populations*
An effective method of studying population growth with micro-organisms is to start a culture each day and, on the final day, to sample and count all cultures. For example, a new yeast culture can be started each day for 10 days, using one

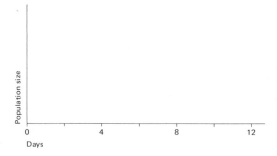

grain of yeast for each culture. On the tenth day samples are taken from each culture and counted with a microscope. A special slide for counting blood cells is desirable, but not essential. If the population on a given day is too large to count, dilute a sample by adding 9 parts water to 1 part of sample (suggestion: use 1 ml of sample and 9 ml of water). The count is then multiplied by 10 to get the actual sample size. If one dilution is not enough, further dilutions may be performed until it becomes easy to count the number of organisms. The multiplication factor for two dilutions is 10 by 10 or 100: for three dilutions it is 10 by 10 by 10 or 1,000. Note that each successive dilution is started from part of the previous dilution, not from the original sample. The data obtained from cultures are graphed for analysis by pupils (see format below, left). Time is the independent variable and population size is the dependent variable.

3.28 Graphing population changes

Let pupils combine or average the data derived from a ten-day population growth study (see experiment 3.27) and graph the results for the entire class. (For example, note that the two-day-old culture was started on the eighth day).

3.29 Human population growth

Ask pupils to compare the results obtained with yeast populations with a curve of human population growth (see graph).

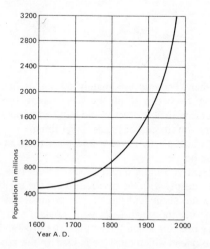

Actual growth of human population of the earth

3.30. Growth of fruitfly population

If a microscope is not available for yeast-cell counting, compare daily counts of fruitflies or some other available population that grows rapidly (see graph).

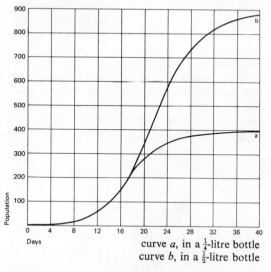

curve a, in a ¼-litre bottle
curve b, in a ½-litre bottle

3.31 Sampling a brine shrimp population

An easily cultured and inexpensive organism for population study is the brine shrimp, *Artemia* sp. Eggs are usually available at tropical fish stores. The eggs will hatch at 21° C in about 2 days when sprinkled on the surface of a salt solution containing 100 g dissolved in 1 litre of water (use non-iodized sodium chloride). Daily population size can be determined by using sampling techniques. Pupils can calibrate a medicine dropper for this purpose by counting the number of drops required to fill a graduated cylinder to the 10-ml mark (e.g. if 160 drops are required, then each drop contains 10 ml ÷ 160 = 0.07 ml per drop). One drop from this medicine dropper is placed on a slide, and the organisms are counted; pupils can then calculate the number of brine shrimp in a known volume of the culture. A graph of daily population changes will give a striking picture of the rate and percentage of hatches from a known quantity of eggs. Egg counting requires a hand lens and graph paper. Spread the eggs as evenly as possible over the paper, and count the eggs in a random sample of squares. Multiply this count by the total number of grids to get the total estimate of eggs.

3.32 *The behaviour of earthworms (Lumbricus)*
A wooden box 30 cm by 30 cm by 15 cm fitted with a glass front is useful for studying the habits of earthworms. Fill the box nearly to the top with successive layers of (a) sand, (b) leaf mould, and (c) loam, padding down each layer before adding the next (see sketch). Place lettuce leaves, dead leaves, carrot, etc., on the surface soil, together with some worms. Keep the contents damp and study the behaviour of the worms.

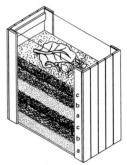

3.33 *An observation nest for ants*
A. Make a wooden U from three 30 cm lengths of wood 1.5 cm square. Fix this vertically on a convenient wooden base. Now cut two rectangles of glass, 30 cm by 33 cm, and clamp them on each side of the U with rubber bands or some sort of metal clip. Make a well-fitting wooden lid, as

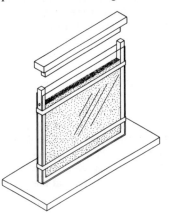

shown in the diagram. Drill a 0.5 cm hole about 5 cm from the top of one of the sides and plug it with cotton wool.

To set up the nest, first fill it with soil taken from the field from which you obtained the ants. Pour sandy soil in the top and pat it down occasionally with a ruler until it is about level with the plugged hole.

B. Now for the ants; small black or red ants are the best for this purpose. They prepare their colonies under flat stones. Raise a flat stone and you will see the ants scurrying away. You will need two narrow-necked medicine bottles with cotton-wool stoppers, a gardening trowel, an insect collector, and a white sheet or large piece of paper.

The insect collector is made from a 25 – 30 cm length of plastic or glass tubing, 2 – 3 cm in diameter, a small piece of fine wire mesh, and a length of flexible tubing (see diagram below). Cut a circular piece of wire mesh slightly larger than the inside diameter of the glass or plastic tubing. Force the screen half-way through the tubing by pushing it with a rod. An alternative is to cut the tubing in half, insert the wire mesh and tape the tubing back

together. Tape approximately 50 cm of flexible tubing to one end of the transparent tubing. Insects are easily collected and handled by sucking sharply on the end of the flexible tubing.

Collect about 100 ants and place them in one of the bottles.

Next you must find a queen. To do this, dig rather deeply with the trowel and put the earth on the white sheet laid flat on the ground. As you break up the earth with your fingers, you will notice one ant much larger than the others. This is the queen which must be guided to the second bottle: some patience is required here.

C. To get the ants into the observation nest, fill a

large tray with water and place an upturned dinner plate in the middle to form an island from which the ants cannot escape. Place the observation nest on the plate and release the ants either on to the plate or straight into the top of the formicarium: once the queen is inside, the others will follow through the hole in the side. Ants do not like daylight, so plug up the hole, fit a brown paper envelope over the case and remove the nest to its permanent home. A little honey smeared on the glass just inside the entry hole will provide plenty of food, and an occasional sprinkle of water with a fountain pen filler will keep the soil moist.

The exciting happenings inside the nest, the laying of eggs, the grubs and the way the ants talk to each other by rapping each other on the head with their antennae, can all be studied in artificial light, which does not disturb them. Since the tunnels must run parallel to the glass, these things are quite easily seen.

Experiments such as the removal and subsequent return of a few ants, the introduction of foreign ants, greenflies, spiders, etc., are all most fruitful.

Once the nest is settled and the queen begins laying eggs, the cotton wool plug can be removed from the hole. Place the observation nest near a slightly open window and the ants will come and go freely for a whole year.

Studying communities

A grouping of populations in a particular location is called a community. Typically, communities consist of plants and animal populations that perform certain roles. Some populations are the producers. They are so called because they have the capability of trapping energy from sunlight and producing food. Populations that feed on other living populations are called consumers. Those populations that feed on dead material are called reducers, since they disorganize organic matter to yield simpler chemical substances.

3.34 *A closed community*
An interesting way to introduce the study of a

natural community is to prepare a model community in the classroom. Let pupils establish several communities that are closed systems to everything except light. Each consists of a jar containing water (without chlorine), a few small fish such as guppies, some water plants and a few snails. Close the jar with the cap and seal the cap by melting wax around the opening between the cap and bottle. Now submerge the bottle in a large glass vessel such as a battery jar filled with water (see diagram). This will show that the system is not open to air. Place the model community on a

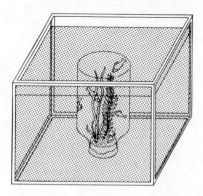

window-sill and have pupils observe it daily. Discuss the relationships between the organisms in the jar. Pupils may wish to attempt to balance a community so that the organisms will survive for a longer time period.

3.35 *Succession in a community*
A hay-infusion culture is an excellent device for demonstrating that a community is dynamic and may change considerably with time. Half-fill a 5-litre jar with dead grass. Cover the grass with water that has been boiled and cooled. Cover the top with a glass plate, cardboard or plywood. Get pupils to examine the jar daily with the naked eye, then with a hand lens, as well as examining samples under the microscope. At first pupils will see bacteria; later, ciliated protozoa will appear (see overleaf). Later rotifers, small round worms (nematodes) and crustacea may appear. Pupils should note the disappearance of populations as

well as the appearance of new ones in the model
community. They should be encouraged to compare gross changes observed with the naked eye
to changes detected by the microscope.

3.35 Protozoa: (a) *Amoeba*; (b) *Parmecium*;
      (c) *Stylonychia*; (d) *Vorticella*; (e) *Colpidium*;
      (f) *Tetrahymena*

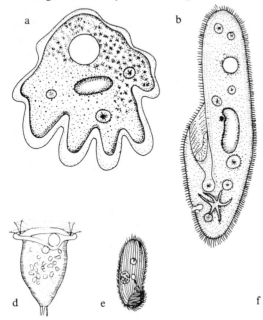

3.36 *A rotting log community*

Break open a rotting log with a trowel, put two
or three chunks into a plastic bag, and take them
back to put in your terrarium. Terraria may be
constructed from aquaria covered with hardware
cloth. If aquaria are not available, glass sheets
may be taped together and placed in a low, flat
waterproof pan. No soil is needed (see sketch). If
the log was in a damp spot, you should add water
to the terrarium from time to time.

Many creatures may live in the log, including
ants, termites, spiders, and horned beetles. If your
log contains some ants, provide a few crumbs and
some sugar water on a piece of sponge for them.
To keep the ants from crawling out of the terrarium, spread a layer of vaseline along the upper
edge. Watch to see what kinds of insects and
other animals come from the log. Some may be in
the form of eggs when you collect the log and may
develop into adults while in the terrarium.

3.36 A rotting log
       community

3.37 *A desert community*

If you don't live near a desert, you'll have to get
material for your desert community from places
near at hand. You can get sand from a beach or

garden supply store. Some kinds of desert animals, including horned lizards, can be found in pet shops. The lizards will eat small insects such as ants and mealworms (also available from pet shops).

You can buy small cacti from florist shops or general stores. Also get some succulents, which are plants that hold water in their fleshy leaves. Besides the plants, put some rocks in the terrarium, making cliffs or overhangs near the edges (see sketch). Put a small dish of water in one corner. Leave an open area of sand in the centre, especially if you have a horned lizard (you will discover why). The temperature of the desert terrarium should be kept between 20° and 27° C.

3.38 *A meadow community*
The problem here is limiting yourself to a few of the many grasses, weeds, seedling trees, and other plants that grow in meadows. There are also many animals to choose from, including spiders that spin beautiful orb webs. These spiders need lots of room, such as a 50-litre aquarium tank, in which to make their webs.

You may find plants with insect eggs or cocoons on them; watch to see what hatches from them. If you want to have one larger animal in the terrarium, try to find a small garter snake. It will eat earthworms and large insects. Be sure to keep the terrarium fairly dry, since snakes often get skin diseases if kept in damp surroundings (see sketch).

3.39 *A forest floor community*
This is the kind of habitat most often modelled in a terrarium. For plants, obtain small ferns, tree seedlings, wildflowers, and especially evergreen plants such as partridgeberry or wintergreen. Put a few of these plants into the soil and cover the rest of the surface with mosses, attractives stones, and perhaps a small limb (see sketch).

For animals, look for small toads, frogs such as cricket frogs or tree frogs, and red newts (which are small salamanders). These animals and the plants of the forest floor all need moisture, so keep the terrarium well-watered and make a small woodland pool in one corner.

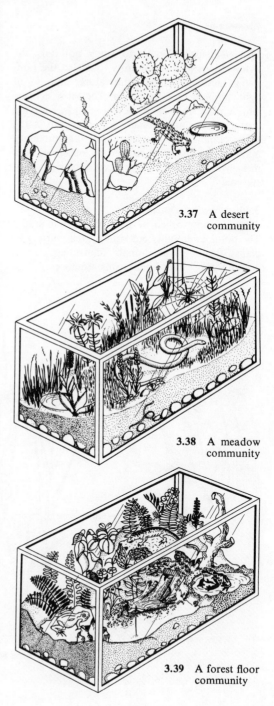

3.37 A desert community

3.38 A meadow community

3.39 A forest floor community

Ecosystems

Biologists usually wish to study not only the
living things in a community, but also the non-
living factors such as temperature, amount of
light, amount of oxygen, etc. The term 'ecosystem' is used when both the living and non-living
characteristics are studied. An ecosystem is the
living community plus the non-living surroundings. An ecosystem is studied by observing and
measuring relationships between its various sub-
systems.

3.40 *A pond ecosystem*

A pond is an excellent subject for pupils to study.
The pond community contains a great variety of
plants (producers), animals (consumers), and
decomposing micro-organisms (reducers) (see
sketch). Observation of feeding habits leads to
an understanding of the food chain in the ecosystem. However, a more quantitative approach is
to dissect collected organisms and examine stomach contents. This, of course, destroys the organisms and hence drastically affects the ecosystem.
It may be preferable to have pupils concentrate
upon gathering information from an ecosystem
in a way least likely to alter or destroy the ecosystem being studied. This will probably necessitate
the use of inferences in place of direct observations. Care must be taken that inferences are not
treated as observations in such cases. For example, the presence of a frog and a bee in the pond
ecosystem may lead a pupil to indicate that a link
on the food chain is bee to frog. However, the bee
may not be eaten by frogs and thus never appear
in the study of frog stomach contents.

Cross-section of a pond, an ecosystem
showing typical life forms

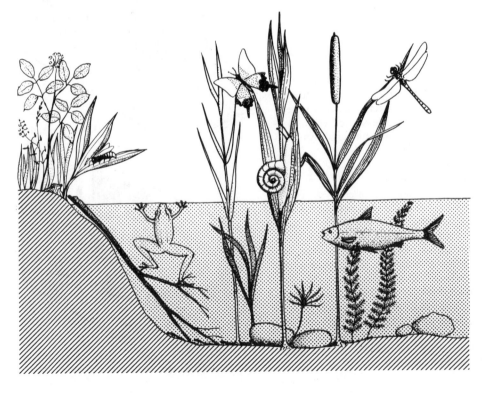

Studying plants

3.41 *Photosynthesis*

The photosynthesis activity of leaves can be demonstrated by placing water plants such as elodea (*Anacharis* sp.) in a funnel, inverting the funnel in a large beaker of water and placing a test-tube over the small end of the funnel (see diagram). A fine piece of tubing or plastic electrical insulation is used in the manner of a drinking straw to remove the air from the test tube, thus filling it with water. Several dabs of putty placed between the funnel and the beaker will permit free circulation of the water from beaker into funnel. Be sure that the water plants have not been in contact with a zinc container prior to placement in the apparatus. Bright sunlight or an electric lamp may provide the light which is required for

photosynthesis. The gas which bubbles from the plant and collects in the test-tube can be tested by thrusting in a glowing wood splint and watching for it to flame. This is the simplest test for oxygen gas. (Since elodea has a hollow stem, punching it at the end with a pin will release oxygen bubbles faster and in a stream so they can be counted to give quantitative results).

3.42 *Respiration*

The respiratory activity of organisms can be demonstrated with an apparatus that moves air over leaves, insects or a small animal and bubbles it through a weak solution of limewater, $Ca(OH)_2$. The system must be isolated from atmospheric carbon dioxide, as indicated.

Set up the apparatus as shown in the diagram, but leave the third bottle empty. Run the appa-

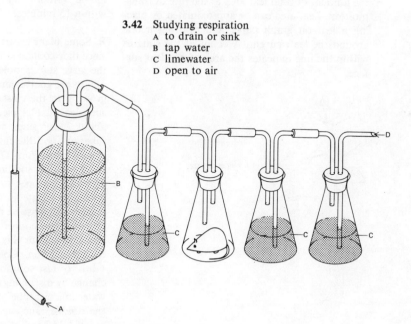

3.42 Studying respiration
A to drain or sink
B tap water
C limewater
D open to air

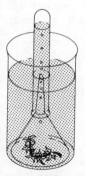

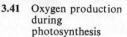

3.41 Oxygen production
during
photosynthesis

ratus by siphoning the large container (a carboy) until it is empty. Note the results. Replace the solutions in all bottles, and this time put loosely packed leaves or an animal into the third bottle. Compare the results of the first run (control) with the second.

Lime water turns from clear to cloudy in the presence of carbon dioxide. This can be demonstrated by blowing through a drinking straw into a container of clear lime water. Pupils can see that plant leaves produce oxygen in some situations and carbon dioxide in others, and that plant leaves produce the same gas as humans under certain conditions.

3.43 *Transpiration*
Leaves also give off water vapour. This can be demonstrated with a potometer (see diagram). Pupils can measure the rate of water loss (transpiration) under differing conditions of humidity, wind and temperature. They can also compare the amount of total leaf area to the rate of transpiration. Leaf area can be approximated by placing a leaf on graph paper and drawing a line around the leaf margin. A count of the squares within the line indicates the area of one leaf surface.

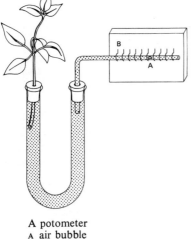

A potometer
A air bubble
B scale

3.44 *The products of leaf activity*
A. Sugar, the product of photosynthesis, and the large starch molecules which are formed from a number of sugar molecules are present in leaves. A simple starch test consists of applying a dilute iodine solution and watching for the typical blue-black colour which indicates that starch is present. The iodine solution is prepared by dissolving 10 g of potassium iodine in 100 cm³ of distilled water and adding 5 g of iodine. Tubers such as potatoes or a starch paste may be used to demonstrate the colour change. When testing leaves it is necessary to soften the leaf cells by boiling in water for a few minutes. Then the leaf is placed in boiling alcohol until the pigments which will mask the reactions are removed from the leaf.

Caution: heat alcohol with an electric heater or use a water bath if a flame is used. Chlorophyll is usually removed in 5–8 minutes but fleshy leaves may take longer or require a change of alcohol for adequate removal of pigments. The iodine solution should react with the starch within 15 minutes.

B. Some plant leaves are suitable for sugar tests, since they contain stored simple sugar rather than the large starch molecule (corn, sugar beets and onions are good examples). Sprouting onion bulbs growing in the classroom are a good source of such leaves. Cut pieces 2 cm long and place in about 2 cm³ sugar test solution in a Pyrex test tube and boil the mixture. (The sugar test solution is made from 173 g of sodium citrate, 200 g of crystalline sodium carbonate, and 17.3 g of crystalline copper sulphate. Dissolve the carbonate and citrate in 100 cm³ water. These substances will dissolve faster if the water is warmed. Dissolve the copper sulphate in 100 cm³ water and slowly pour this solution into the carbonate citrate solution. Cool and add water to make 1 litre of test solution.) Demonstrate the colour change by dissolving a little cane sugar in 10 cm³ water in a test tube. Add saliva which will change the double cane sugar into a simple sugar (glucose). Add 3 cm³ of the test solution and boil over a heat source. A yellowish or reddish precipitate forms when simple sugar is present.

3.45 *Measuring leaf activity*

Brom-thymol blue solution is used to indicate the presence of carbon dioxide. Fill four test-tubes three-quarters full of water. Add approximately 25 drops of brom-thymol blue to each tube. Place a sprig of elodea or other small water plant in two of the tubes (see diagram). With a drinking straw, blow bubbles into one of the tubes not containing a plant, and then into one with a plant. Note the colour change which indicates the presence of carbon dioxide. Now stopper all four tubes and observe the changes within 15 minutes to an hour. Repeat the experiment, but place the tubes in a dark place such as a closed desk.

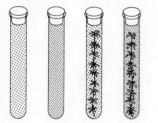

3.46 *Growing plants in the classroom without soil*

A. A sweet potato will produce dense foliage in the classroom if it is placed in water. Set the potato, root end down, in a glass or jar and keep the lower third covered with water. The potato may be kept in position by pressing three tooth-picks or matches into its side and resting them on the rim of the jar (see sketch).

B. The roots of carrots, beets and turnips contain much stored food. They will produce foliage if grown in water but will not develop into new plants. Remove the old leaves from the top and then cut off all the root except 5 to 8 cm. Place this portion in a shallow dish of water. A few pebbles placed in the dish will hold it upright.

C. Cut off a pineapple 3 to 5 cm below the base of the leaves, and set this portion in a shallow dish of water. The leaves will continue to grow for several weeks.

3.47 *Osmosis*

Select a carrot or potato which has a large top and which is free of breaks in its surface. With a sharp knife, an apple corer, or cork borer cut a hole in the top of the carrot about 5 cm in depth. Be careful not to split the top. Fill the cavity with a concentrated solution of sugar. Insert a tightly fitting one-hole rubber stopper which carries two soda straws pushed together or a length of glass tubing (see diagram). Place in a jar of water for a few hours. If your cut in the top of the carrot has not been even, it may be necessary to seal the cork in with some wax dripped from a burning candle. Pupils enjoy competing to see who can attain the greatest height of water in the tube.

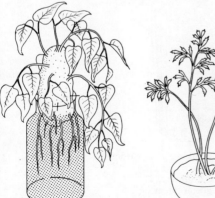

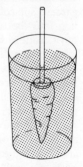

3.46 Plants in the classroom

3.47 Osmosis

3.48 *Plant parts will grow roots*
Obtain a box of sand and place it away from
direct sunlight. Wet the sand thoroughly and
keep it moist. Plant any of the following in the
sand: (a) various bulbs; (b) cuttings of begonia
and geranium stems; (c) a section of sugar-cane
stem with a joint buried in the sand; (d) a section
of bamboo stem with a joint buried in the sand;
(e) carrot, radish and beet tops, each with a small
piece of root attached; (f) an onion; (g) an iris
stem; (h) pieces of potato containing eyes; (i) a
branch of willow.

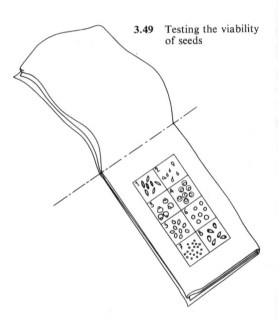

3.49   Testing the viability
of seeds

3.49 *Testing for root growth from seeds*
Fold a square metre of muslin twice in the same
direction. Near one end mark out with a pencil
8 or 10 squares about 5 cm by 5 cm (according to
the number of kinds of seeds available). Number
the squares, and on each square place ten seeds
from a particular packet (see diagram). Note the
arrangement of the different seeds. Fold the
opposite end of the muslin over the seeds. Roll
up the tester and tie it loosely with string. Saturate the tester with water. Keep it moist and in
a warm place for several days. Then unroll it and
see how many of each kind of seeds have germi-
nated. Have pupils report the viability as a percentage or by constructing graphs.

3.50 *A drinking-glass garden*
Each pupil should make his own drinking-glass
garden. Slip rolled blotting paper or paper towelling into the tumbler. Fill the centre with peat
moss, cotton, sawdust, or wood shavings. Slip a
piece of graph paper with water-insoluble ink,
cut to size, between the glass and the paper.
Plant bean seeds between the paper and the
glass, as shown in the diagram. Water the centre
of the tumbler. Have pupils make frequent observations of growth, using the graph paper lines
as reference points. Results may be presented by
constructing graphs which portray time on the
horizontal axis and the root and stem elonga-
tion in millimetres along the vertical axis.

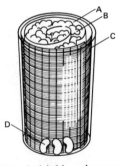

3.50   A drinking glass garden
  A paper towelling
  B cotton, etc.
  C graph paper
  D seeds

3.51 *Germinating pollen*

Make a strong sugar solution and place it in a shallow dish like a saucer. Shake pollen from several kinds of flowers on to the surface of the sugar solution. Cover with a sheet of glass and let it stand in a warm place for several hours. If the experiment is successful you will be able to see little tubes growing from the pollen grains. Use a hand lens.

3.52 *Seed structures*

Soak seeds of bean, pea, pumpkin, sunflower, peanuts, corn and other large forms. Remove the seedcoats and carefully cut the seeds open. Discover the parts that make up the seed. There is little point in teaching the botanical names of these parts, though pupils may enjoy learning them. It is of more importance that pupils learn to recognize the part of a seed that is the young plant and the part that is stored food (see diagrams).

Comparative structures of a dicotyledon (bean) and a monocotyledon (corn). Note also that in seeds such as beans the embryo absorbs all the endosperm, but in a species such as corn the embryo does not absorb much of the endosperm until the seed germinates.

A seed coats
B embryonic shoots
C endosperm
D cotyledon
E embryonic root
F epicotyl
G hypocotyl

Germination and early development of a bean plant
A leaf
B stem
C cotyledon
D primary root
E seed root

3.53 *The parts of a flower*

Examine specimens of large simple flowers such as tulips or lilies. Count the stamens and observe how they are arranged about the central pistil. Make large diagrams of the essential organs. Label the parts of the pistil (stigma, style and ovary). Label the parts of the stamen (filament and anther).

The end of the stalk on which the flower grows is called the receptacle. At the base of the receptacle there are usually leaf-like structures that enclose the bud. These are called sepals. Above the sepals there is usually a ring of brightly coloured petals called the corolla (see sketch).

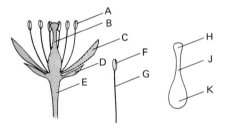

The parts of a flower
A stamens
B pistil
C petals
D sepals
E receptacle
F anther
G filament
H stigma
J style
K ovary

3.54 *Dissecting flowers*

A. If a sufficient number of flowers are available, this exercise is most valuable as an individual pupil activity. Simple flowers with a single row of petals should be selected. Label each of five cards or pieces of paper with one of the following words: stamens, pistil, petals, sepals, receptacle. Dissect a flower carefully and place the parts neatly on the appropriate cards. Some flowers can be pulled apart quite easily, but a knife or scissors may be needed for others.

B. Pick up one of the stamens and rub the anther lightly across a piece of black paper. Traces of pollen will usually be seen. Observe the pollen with a microscope.

C. Cut the ovary crosswise with a sharp knife and count the ovules or 'seed pockets'. Look for traces of seeds in the ovules.

3.55 *Fruits development*

A. Collect specimens of flowers in different stages of maturity, from newly opened buds to specimens in which the petals have fallen. Cut each ovary open, and note the changes that take place during seed development. Roses, apples and tomatoes are good for this purpose.

B. Look over 1 kg of fresh-picked peas or string beans or other legumes and pick out the pods that are not completely filled. Open these and compare them with fully filled specimens. The abortive seeds are the remains of ovules that were not fertilized by pollen.

3.56 *Monocots*

Obtain several stems of several plants such as bamboo, sugar cane and corn. Cut each of the stems crosswise with a very sharp knife or razor blade. Observe the similarities in the cross-sections. Especially notice that the tubes of fibrovascular bundles are scattered throughout the pith on the inside of the stem.

3.57 *Dicots*

Obtain the stems of several plants or small trees such as willow, geranium, tomato, etc. Cut across each of these stems with a sharp knife or razor blade. Observe that just under the outside layer of the stem there is a bright-green layer. This is the cambium layer. Also observe that the tubes of fibrovascular bundles are arranged in a ring about the central, or woody, portion of the stem.

3.58 *Light affects stems*

A. Plant some seeds that grow rapidly such as oats, radish, bean or mustard seeds in two flower pots. When the seedlings are about 2.5 cm high, cover one pot with a box that has a hole cut near the top. From time to time lift the box and observe the direction of growth. Turn the box so that light comes from a different direction and observe again after a few days.

B. Put two light baffles in a long, narrow box as shown in the diagram, and cut a hole in the end. Plant a sprouting potato in a small pot that will

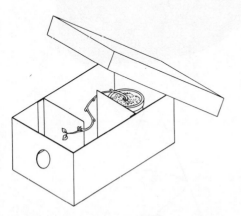

fit in the box. Place the pot behind the baffle farthest from the hole. Cover the box and place near a window. Observe the direction of growth from time to time.

C. Plant some fast-growing seeds in four flower pots. Keep the pots in a darkened room until the seedlings are about 2.5 cm high. Place one pot near a sunny window and observe the effect. Turn the plants away from the light and observe. Leave the pot in a place away from direct light for a few days and observe the results.

D. Place each of the three remaining pots of seedlings in a different box. Cut a window in each box and cover each window with a different colour of cellophane such as red, yellow and blue. Place the three boxes containing the pots of seedlings in good light with the window facing the light. Observe any difference in the effect produced by different coloured light on the growth of stems.

3.59 *Gravity affects the growth of stems and roots*

A. Place some seeds that grow rapidly such as oats, radish, or mustard seeds on moist blotting paper between two glass plates secured with rubber bands. After germination, turn the apparatus vertically through 90° and allow to remain undisturbed. Repeat the turning at intervals and observe the effect on the roots.

B. Another way to study the effect of gravity on roots is to sprout some seeds and select one that is straight. Pierce the seed with a long pin or needle and stick this into a cork. Place some damp cotton or blotting paper in a bottle. Put the cork and seedling in the bottle (see diagram). Place the bottle in a dark cupboard and look at it every four hours.

Gravity affects roots

3.60 *Studying stem tissues*

Stem cross-sections are excellent for use with a micro-projector or low-power microscope (see overleaf). Stem cross-sections are easy to slice thin enough for microscopic viewing. A comparison of monocotyledon stems with dicotyledonous plant stems provides an interesting beginning to the study of vascular plants. Placing the stem in a beaker of water with food colouring or red ink added may provide an indication of those cells involved in the upward movement of water.

Celery or beans are also used for this purpose. Upward movement of water in cut stems is enhanced when the final cut is made under the water solution. This prevents the formation of air bubbles which inhibit the flow of water up the conducting vessels.

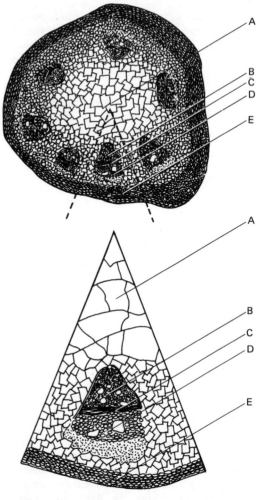

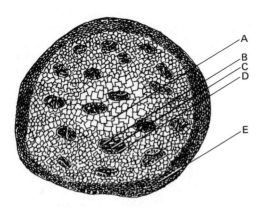

Stem cross-sections
A pith
B xylem
C cambrium
D phloem
E epidermis

Studying animals

3.61 *Activities occurring in animal organs*
The absorption of carbon dioxide by a potassium hydroxide solution provides a means for measuring the oxygen intake of a small animal. Place a cockroach, a locust, or several mealworms in a closed container in which paper soaked in 0.5 per cent potassium hydroxide solution is suspended. The two-hole stopper should contain a 0.2-ml pipette or a glass tube of fine bore (see

opposite page). If glass tubing is used, you will need to place a piece of 1-mm graph paper or ruler behind it for reference points to observe the movement of a drop of coloured water or dye through the tube. Keep the organism off the absorbent paper by adding crumpled paper or tying the moist paper on a string attached to a drawing pin stuck in the underside of the stopper. You should have pupils devise a control jar which contains everything except the organism. Record the movement of the drop of dye at regular intervals. Compare with the control. The difference is due to oxygen conversion to carbon dioxide in the organism. Let pupils examine the organism for structures that are involved in air entrance such as spiracles.

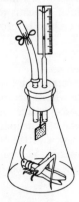

Measuring the oxygen
intake of a small animal

3.62 *Observing a snail heart*
If you have an aquarium with snails, look for thin gelatinous snail masses attached to the glass or on leaves. These can be removed intact with a razor blade and placed on a microscope slide containing a few drops of the aquarium water. Observe with a microprojector or low-power microscope. The beating heart is easily seen.

Studying tissues

3.63 *What is a tissue?*
Groups of similar cells performing a similar function within a multicellular organism are called tissues. The activity of an organism frequently involves the co-ordination of several different tissues. A good illustration of interrelated tissues is the dissected leg of a chicken. Pupils can see how toe movements are dependent upon the various tissues that make up the leg. The movement of toe bones can be controlled by pulling specific tendons. To prepare the leg, remove the skin and separate the tendons by

removing connective tissue down to the toes. Get pupils to discover which tendons (and hence which muscles) pull toes up and which pull them down. Discuss the functions of the bone, tendons, muscles, blood, blood vessels, and nerves found in the chicken-leg preparation.

3.64 *A liquid tissue*
Blood is a convenient tissue for study. It has a great variety of interesting and unique characteristics such as clotting action, presence of antibodies, and blood types that make it an excellent introduction to a study of organ transplants, genetics, respiration and a variety of other topics. In general, the use of blood samples taken from pupils should be avoided. Frog blood, mammalian blood from a slaughterhouse or butcher shop or outdated whole blood from a hospital are common sources.

Caution: if it is proposed to use samples taken from pupils, the teacher should first check the local school and medical regulations, which often prohibit this practice. If approval is granted, make it a rule that a lancet must never be re-used, owing to the possibility of the spread of infections such as hepatitis. Flaming a needle or dipping in alcohol will not prevent such transmissions.

Mammalian blood can be kept from clotting by adding a 2 per cent sodium citrate solution to the blood in a ratio of 1 part solution to 4 parts of blood. Let pupils observe frog and mammalian blood with a microscope for a comparison of blood cells with or without nuclei.

3.65 *Observing blood flow*
Blood cells may be observed in certain living organisms such as fish or frogs. The fish or frog can be wrapped in a wet cloth and pinned to a soft board with a hole for observing with a microscope. The fish tail-fin, or frog webbing between the toes, is mounted over the hole so that the entire preparation can be placed on a microscope stage (see overleaf). Blood cells can be seen squeezing through the extremely small blood vessels of the thin fin or web. Another interesting liquid tissue is the milk of a coconut. This rich nutrient

Fish and frog prepared
for observation of
blood flow under a
compound microscope

is often used in tissue cultures as a food source for growing tissues from one or a very few individual cells.

Studying cells

3.66 *What is a cell?*

A great variety of cell types are available for microscopic examination in the science classroom. Although a few cell types are relatively large (e.g. an ostrich egg or some marine algae cells), most require a microscope for efficient study. There are two potential sources of cells for study. There are cells which are considered whole organisms, called protists. Yeast cells, protozoa, bacteria, euglena and other one-celled organisms are examples. Since groups of these one-celled organisms are actually a population, it may be preferable to begin cell study by using the cells from the tissues of multicellular organisms. A striking comparison of the cells from animal tissue and plants is easily arranged. A plant that

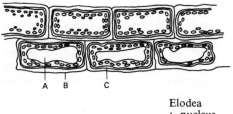

Elodea
A nucleus
B cell membrane
C cytoplasm

is very effective for this purpose is commonly found in aquaria; it is called elodea (*Anacharis* sp.) (see diagram). The small leaves near the termination of a stem are best. Place a single small leaf in a drop of water on a glass microscope slide, cover with a cover slip and examine with a microscope. In strong light the cellular contents may exhibit a flowing motion called cyclosis or protoplasmic streaming.

3.67 *Plant and animal cell differences*

With a clean toothpick, gently scrape the inside surface of your cheek. Place the whitish scraping into a drop of water on a microscope slide. Add a drop of stain such as methylene blue or iodine, and cover with a cover slip. View with the low, then high, power of the microscope. Let pupils compare the animal cell with the plant cell (see sketch).

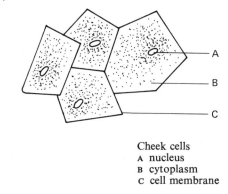

Cheek cells
A nucleus
B cytoplasm
C cell membrane

3.68 *Cell walls*

The elodea slide (see 3.66) may be used to demonstrate the presence of a cell wall. Place a drop of salt-water solution on one edge of the cover slip. Draw the salt solution under the cover slip by placing a piece of absorbent towelling or blotter at the opposite side of the cover slip so that the liquid on the slide will rise up the paper. Water will diffuse out of the cells into the salt water. As this proceeds the cellular contents may be observed to shrink, but the rigid cell walls retain their original structure. Other plant cells may be used to demonstrate this phenomenon. Fleshy leaves that have a thin layer which can be peeled off are possible sources of thin cellular layers. *Trades-cantia*, lettuce and spinach are examples.

Elodea cell in salt solution

3.69 *Cell reproduction*

The fundamental reproductive process known as cell division may be studied by selecting an appropriate tissue that is growing rapidly. A good source of such cells is the root tip region of onions

or other related plants. Onion bulbs, garlic cloves or onion sets placed in an aerated water bath provide large quantities of material. Cut off the white root tips of healthy specimens. Cut a 3-mm cylinder from the end of a root. Place it in a drop of aceto-carmine stain on a microscope slide. Cut up the onion tip with a razor blade until the pieces are extremely small. Cover the preparation with a cover glass. With a piece of folded paper towelling over your thumb for protection, gently squash the pieces of root tip by pressing on the cover slip with a rolling motion. Do not allow the cover glass to slide. Then examine the

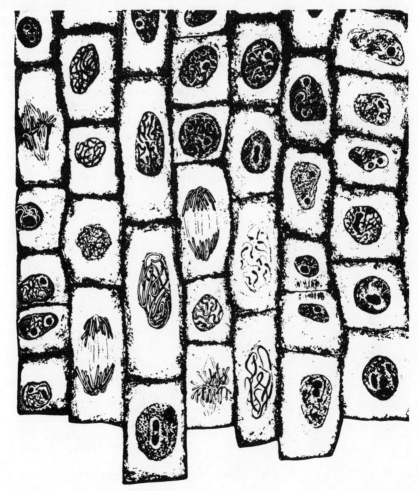

Cell division

preparation with the low power of a microscope. Look for dark-stained threadlike bodies. These are chromosomes or mitotic figures. Let pupils find various types or stages. After perfection of the technique, let pupils count the number of various stages and from this information have them approximate the relative lengths of time that the various stages are present in a reproducing cell (see illustration on previous page).

3.70 *Salivary chromosomes*

The large size of the chromosomes in salivary gland chromosomes of fly larvae make it easy to study these cell organelles. The larvae of Sarcophagid flies, called blowflies, can be collected from uncooked meat left outdoors. The salivary glands are exposed by placing a larva in a drop of salt solution on a slide. With two dissecting needles, hold the tail end with one needle and pierce the head with the other needle. Slowly stretch the larva until the head, mouth parts, digestive tube and salivary glands are pulled free. Separate the fat cells, digestive tube and head from the salivary glands. Under a microscope, a cell squash of the glands stained with aceto-orcein stain should reveal large banded chromosomes.

3.71 *Observing organelles*

In recent years the development of the electron microscope has provided information about the structures that are contained within cells. These cell parts are called organelles. Although pupils cannot usually see or visit an electron microscope facility, they should be aware of its importance in extending the biologist's understanding of the levels of organization below that of the cell.

Photographs of electron microscopes are readily available from manufacturers of the instrument. Also examples of the working photos produced by this instrument are provided free by some manufacturers, as well as certain pharmaceutical companies. Such photos with accompanying class discussions enable pupils to begin visualizing the structures and places in cells that are the sites of specific activities (see diagram). For example, the process of respiration is less mysterious when a student sees photos of the structures called mitochrondria, where the process takes place.

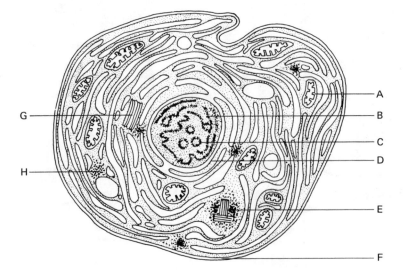

3.71 A typical cell
  A mitochondria
  B chromosomes
  C nucleolus
  D nuclear membrane
  E centrioles
  F cell membrane
  G Golgi bodies
  H ribosomes

Chapter Four

Earth and space sciences

Introduction

Earth science topics have always been of interest to children. This interest ranges from amazement at the beauty in a so-called 'pretty' rock to 'Whatever happened to dinosaurs?' Recent events in space science have made this topic of equal interest. It is not necessary that the instructor be either a geologist or a space astronaut to teach these topics to elementary school children.

The teacher should not feel it necessary to be able to answer every question. Much earth science can be learned without the use of difficult vocabulary or sophisticated concepts. It is not the purpose of the material on the following pages to train junior geologists but to promote in children an interest in earth science.

Rocks and minerals

Simple beginnings

4.1 *Where to start*

A *Essential materials for collecting, identifying and mounting rock and mineral samples*
Specimens of igneous, metamorphic, and sedimentary rocks;
Collection bag or knapsack, paper, pencils and rubber bands;
Magnifier (10 power);
Hammer;
Eye-dropper and vinegar and/or dilute HCl;
Nail varnish for writing names or numbers on rocks;
Glue for mounting rock samples on show-cards;
Test-tube clamps, and test-tubes;
Chisel;
Bronze coin;
File;
Porcelain streak plate or tile;
Small weighing scale;
Magnet;
Alcohol lamp;
Boxes for separating rocks into groups;
7.5 × 12.5 cm cards for recording data;
Graph paper (millimetre squares);
Small glass jars;
Glass plate;
Tweezers.

B *Rock and mineral observations*
Get each pupil to bring in a rock or several rocks that for one reason or another may have some appeal to him. What kinds of questions can the pupils ask about this rock? For example, How old is it? How did it get where it was found? Is it valuable? How many uses can you think of for this rock?

All science begins with observation. Using all the senses, have the pupils describe their rock or rocks. What colour is it? Compare its weight with a known object of similar size. Is it relatively hard or soft? Does it break easily? Place two different rocks side by side. Describe similarities and differences. Place all the rocks contributed by the pupils in a pile. Ask the class to separate these into distinct groups based on similarities and differences. This leads towards the simple identification of rocks and minerals.

4.2 *How to look at a rock*
Children do not as a rule bring in exotic mineral samples purchased from commercial suppliers. Usually they bring in so-called road, rock samples. These samples are usually weathered. In collecting samples, instruct your pupils to obtain samples that show fresh surfaces. This may involve some chipping away and even this is not a guarantee, for rocks will break along pre-existing cracks and some alteration may already exist. Hammer the rock hard enough to reveal unaltered surfaces. With a little experience this should not be too difficult. Rocks may be broken safely by wrapping them in a cloth, placing them on a large rock, and striking hard with a hammer. The cloth wrapping will prevent small chips from flying off. Compare the appearance of freshly broken surfaces with the weather-worn outside of the rock.

4.3 *How a rock differs from a mineral*
A *mineral* is a non-living substance that may have a characteristic shape and which has a definite chemical composition throughout. A *rock* is a composite of more than one mineral, so that if the rock is broken, the pieces might be composed of different minerals.

4.4 *The eight most abundant elements in the earth's crust are as follows*

Percentages by weight	Element	Chemical symbol
46.60	Oxygen	O
27.72	Silicon	Si
8.13	Aluminium	Al
5.00	Iron	Fe
3.63	Calcium	Ca
2.83	Sodium	Na
2.59	Potassium	K
2.09	Magnesium	Mg

These elements form compounds which are also called minerals. Geologists have discovered, named, and classified over 2,000 minerals. However, only a few of these make up most of the earth's crust.

Physical properties of a mineral

While a mineralogist has many techniques for identifying more than 2,000 minerals, here we are mainly concerned with the rock-forming minerals essential to rock identification. Thus we will restrict ourselves to the following testing and descriptive techniques which help us identify some basic rock-forming minerals:

Lustre	Transparency
Hardness	Crystal form
Streak	Other special items
Colour	such as taste, odour
Relative density	magnetism and
Rupture (cleavage and fracture)	structure

Definitions of each of these mineral properties are given below.

4.5 *Lustre*
Lustre is the appearance of the surface of a mineral in reflected light. Minerals are divided into two great groups on the basis of their lustre. One group is opaque and has a metallic lustre

like that of a metal. The other group may be opaque or transparent but does not have a metallic lustre.

4.6 *Hardness*
Hardness is the resistance a mineral offers to scratching. Rocks are scaled in hardness within a range from 1 to 10, 1 being very soft and 10 being extremely hard. To determine the hardness of a sample, hold a grain of the material in tweezers and try to scratch your fingernails with it. If it doesn't scratch your fingernail, the specimen is softer than your fingernail. A fingernail has a hardness of 2.5; therefore, the specimen must have a hardness of less than 2.5. Do the same thing with a piece of copper. This has a hardness of 3. If the specimen scratches the copper, it is harder than it and has a hardness greater than 3. How much greater than 3, we don't know. We must continue until we find a substance that the specimen will not scratch. A steel knife blade is 5.5 and window glass has a hardness of 5.5 to 6.0. Diamonds have a hardness of 10.0. Diamond is the hardest mineral. In hardness testing where glass is being used, make sure that the pupil does not hold the glass in his hand when attempting to do the scratch test. As a safety measure, place the glass on a flat surface prior to use.

The relative hardness of a mineral can be determined by comparing the specimen to a series of minerals chosen as a hardness scale. The standard scale of hardness is : 1. Talc; 2. Gypsum; 3. Calcite; 4. Fluorite; 5. Apatite; 6. Orthoclase feldspar; 7. Quartz; 8. Topaz; 9. Corundum; 10. Diamond.

4.7 *Streak*
Streak is the colour of the ground-up or powdered mineral. This may be obtained by rubbing the mineral on a ceramic streak plate or by grinding the mineral and observing the colour. The colour of the streak may be similar to the colour of the gross mineral, or it may be quite different. The colour of the streak of a particular mineral is usually constant even though the colour of the mineral may be quite variable. Streak plates may be made or improvised from old building tiles or unglazed (or broken) porcelain.

4.8 *Colour*
Colour is the most obvious physical property of a mineral. Yet, because of its variability, it is not considered a reliable property for identification.

4.9 *Relative density*
The relative density of a mineral is a number which expresses the ratio between its mass and the mass of an equal volume of water at 4° C. If a mineral has a relative density of 2, it means that a given specimen of that mineral has twice as much mass as the same volume of water.

Most common minerals have a relative density of 2.5 to 3.0. Minerals less than 2.5 feel 'light' and those more than 3.0 feel 'heavy' for their relative size.

The relative density of a mineral of fixed composition is constant and its determination is frequently an important aid in identification of the mineral. (See also experiment 2.14.)

Several conditions must be observed in order to determine accurately the relative density of a mineral. First the mineral must be pure—a requirement frequently difficult to fulfil. It must also be compact, with no cracks or cavities within which bubbles or films of air could be imprisoned.

The beam balance is a very convenient and accurate apparatus for the determination of relative density. Owing to its simplicity, it can be easily and inexpensively constructed at home (see Chapter One).

4.10 *Rupture or breakage*
A rupture in which the mineral tends to split along planes parallel to crystal faces and with smooth flat surfaces along these planes is called a cleavage. Some minerals have but one cleavage direction, others may have two, three, or more. Any breakage other than cleavage is called a fracture.

4.11 *Transparency*

Transparency is the degree to which minerals will transmit light. Transparent minerals allow full passage of light, like a window pane. Translucent minerals allow passage of light, but not images. Opaque minerals show no passage of light.

4.12 *Crystal form*

Crystal form is the external shape of mineral which reflects the internal arrangement of atoms. Most minerals are crystalline, and the regular and definite pattern of the internal atomic structure results in a definite external shape. A few minerals are amorphous (non-crystalline).

4.13 *Others*

In addition to the above physical properties of minerals, the following are useful diagnostic tests.

Magnetism. The reaction of the specimen to a magnet is of interest. Is the specimen attracted or not?

Hydrochloric acid test. Does the application of dilute hydrochloric acid cause any reaction? Does it bubble or effervesce?

Basic rock-forming minerals

Few rocks are elements, for example, pure gold or silver. Most rocks are combinations of elements. Quartz, for example, is a combination of the elements silicon and oxygen. It is thus called a mineral. Basically minerals, rather than elements, are the constituents of rock.

4.14 *Quartz*

Quartz is a translucent or transparent mineral with no cleavage. It resembles pieces of glass and may be white, milky, smoky, pink, clear, purple or, rarely, green or brown. It is resistant to weathering, has a hardness of about 7 (relative density 2.65) and occurs in the light-coloured rocks. Because it is resistant to weathering,

quartz is one of the main constituents of weathered products like sandstones and siltstones. It may be distinguished from calcite, another rock-forming mineral, by its hardness and lack of effervescence with cold dilute hydrochloric acid.

4.15 *Feldspars*

The feldspars are pink, white, grey, bluish, and red. When these minerals are found in rocks, flashes of light will reflect from the tiny cleavage surfaces. These cleavage surfaces clearly separate the feldspars from quartz, which has no cleavage. The feldspars exhibit two directions of cleavage approximately at right angles to each other. On one cleavage surface of *plagioclase*, which is usually white, grey, or bluish, there are fine lines (striations) representing separation planes between tabular or sheet-like crystal twins. Observation of the twinning striations is a clear-cut method of identifying plagioclase. *Orthoclase* feldspar which is usually pink, red, or white does not have the twinning striations. Feldspars have hardness 6.0 and relative density 2.4-2.7.

4.16 *Micas*

Mica is a group of rock-forming minerals, the most important of which are *muscovite* and *biotite*. The micas are easy to separate by colour. Muscovite is clear and colourless; biotite is brown or black. Mica can be split into very thin sheets. It is elastic and, if bent, will regain its original shape and size. Its hardness is between 2.0 and 2.5. Relative density: 2.7-3.0.

4.17 *Pyroxenes and amphiboles*

These are two groups of rock-forming minerals. They are identified by cleavage and crystal form but are difficult for the novice to distinguish. Most of these minerals are dark, usually ranging from dark-green to black. The most important amphibole is hornblende.

4.18 *Olivine*

Olivine is green to greenish-brown, and weathers easily so that it leaves the rock brown with iron oxide stain. In pure form, olivine occurs in sugary mineral aggregates. The tiny grains sparkle like quartz; however, quartz and olivine seldom occur naturally together in igneous rock, which is formed from the molten state (experiment 4.21). They can occur together in sedimentary rocks (experiment 4.22). Olivine is native to the darkest rocks, those which are severely deficient in silicon. The hardness of olivine is 6.5-7.0. Relative density: 3.2-3.6.

4.19 *Calcite*
Calcite is a basic rock-forming mineral belonging to the group of carbonates. Its lustre is glassy to dull. It has a hardness of 3.0. It is commonly colourless or white and has a colourless streak. Calcite commonly ruptures, exhibiting cleavage in 3 directions not at right angles, resulting in a characteristic rhombohedral shape. Calcite has a relative density of 2.72. It effervesces readily in cold, dilute, hydrochloric acid.

4.20 *Notes for identification*
In general, the following notes should be valuable in identification:

Quartz. Transparent to translucent, glassy lustre, scratches glass, broken surfaces are curved or smooth.

Mica. Soft, shiny, flat flakes (probably black ... biotite).

Feldspar. White to grey to pink, almost opaque, not quite so hard as quartz, dull surface except when light strikes certain surfaces at just the right angle.

Hornblende. Black, hard, long grains.

Calcite. Effervesces with dilute acid.

Major groups of rocks

There are basically three major rock groups: igneous; sedimentary; and metamorphic.

4.21 *Igneous rocks*
Igneous rocks (fire-formed) solidify from a molten fluid condition—usually referred to as magma—and are either 'squeezed into' sub-surface space

(intrusive rock) or 'squeezed out' on to the surface of the earth (extrusive rock). Thus, we have a magma or molten fluid with a specific chemical composition unique to that particular melt that can be either 'squeezed in' or 'squeezed out'. In either case the basic chemical composition will be similar and the only significant difference will be the texture, which is a term referring to the crystal grain size in rocks. Intrusive rocks are coarse textured and extrusive rocks are fine grained or fine textured. The texture of a rock is a function of the rate of cooling—the more rapidly the 'melt' or molten liquid cools, the finer the texture. If a molten mass cools rapidly owing to its exposure to the air, as in the case of an extrusive rock, we find that the texture is fine. And in the case of an intrusive, or 'squeezed in' rock, where cooling occurs underground and is slow, crystals grow larger and result in a coarse texture.

Texture of igneous rocks

Igneous rocks may be divided into two groups —the light-coloured (silicon-and aluminium-rich) and the dark-coloured (iron-, magnesium- and calcium-rich). There are eight basic constituents of igneous rocks, and a reasonable knowledge of how to recognize these minerals is essential before we can classify rocks. They are:

1. *Light-coloured* (silicon and aluminium): (a) quartz; (b) orthoclase feldspar; (c) plagioclase feldspar; (d) muscovite mica.
2. *Dark-coloured*, basic (iron-, magnesium- and calcium-rich): (a) biotite mica; (b) amphibole (hornblende); (c) pyroxene; (d) olivine.

Igneous rocks are for the most part hard, tough rocks, consisting of intergrown grains of silicate minerals.

    The texture of a rock is the pattern determined by the size, shape, and arrangement of grains composing it. Igneous rocks are characterized by uniformity of texture (see figure, previous page) except for the porphyries, where larger crystals are embedded in a fine-grained ground mass. Some igneous rocks are distinctly granular; others, so fine that no individual grains are visible, are said to be dense; and some are glassy or amorphous.

Generally, the grains of igneous rocks are angular and very irregular because during their growth the mineral particles crowded against each other and became interlocked.

4.22 *Sedimentary rocks*
Sedimentary rocks are made up of material derived from previously existing rocks. Minerals found in a sedimentary rock may include all the minerals from metamorphic rocks, igneous rocks and other sedimentary rocks. Some of these minerals are incorporated into some sedimentary rocks with little or no alteration of the physical or chemical make-up of the minerals. In contrast to this, some minerals undergo severe mechanical weathering before being incorporated into a sedimentary rock (see figure). Weathering may completely destroy certain minerals and reconstitute the resulting chemical material into new minerals. Pupils should be able to identify some of the common sedimentary rocks, such as conglomerate, sandstone, shale, limestone, and chert. An important mineral in sedimentary and metamorphic rocks (but not igneous rocks) is calcite.

ments. The hardness depends on how well the grains are cemented together. These rocks are usually less compact than igneous rocks and if you breathe on them, the added moisture gives the rocks an earthy smell. They are apt to be crumbly.

Sediments consisting of broken particles of the parent rock are called *clastics*, e.g., sandstones. These particles can vary from silt particles (0.004-0.06 mm) to sand grain particles (0.06-2 mm) to pebbles (2-64 mm) up to cobble and boulder size. The cementing agent can be a great variety of things such as silica, calcium carbonate, and iron oxides. The most common minerals in the fragmental rocks are quartz, feldspar, and clay minerals.

Some previously existing rocks can be transformed into sedimentary rocks without any evidence of clastic or fragmental particles by being absorbed into solutions and carried away by running water. Sediments from materials carried in solution are called precipitates, e.g., limestone.

4.23 *Metamorphic rocks*
The minerals found in metamorphic rocks are much the same as those found in igneous and sedimentary rocks, with the exception of a few minerals formed by recrystallization, replacement and high temperature.

Metamorphic rocks are rocks that have been transformed from some previously existing rock, be it igneous, sedimentary, or metamorphic, to a

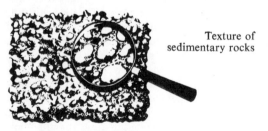

Texture of
sedimentary rocks

Texture of
metamorphic rocks

Sedimentary rock formation involves the breaking up by mechanical and chemical processes, and the moving, sorting, and spreading out of particles from any previously existing rocks, followed by compaction or cementation of sedi-

newly formed rock. Metamorphic rocks are the result of heat and pressure applied to these formations (plus permeating fluids). The texture of a characteristic metamorphic rock is shown in the diagram.

The lining-up of flat mineral grains in a single plane, giving the rock a tendency to split in this direction, is called foliation, and the rock is said to be foliated. Like igneous rocks, metamorphic rocks are hard and tough, and have interlocking or intergrown mineral grains; they differ in that they exhibit foliation. Although foliation is the outstanding characteristic of metamorphic rocks, a few of them, such as marble and quartzite, are not foliated. Three chief varieties of foliation are:

1. *Gneissic or banded:* an imperfect or rough foliation in which the layering is expressed as distinct bands of different minerals. The thicker bands are commonly composed of feldspar.
2. *Schistosic or well-foliated:* the foliation is caused by the parallel arrangement of platy minerals, such as micas.
3. *Slaty cleavage:* the tendency of a rock to split into thin, even slabs such as slate. The minerals are so small that they cannot be seen by the unaided eye, yet the cleavage is the result of the parallel planar arrangement of many microscopic mineral grains.

Metamorphic rocks are classified into two major groups, based on the presence or absence of foliation. The type of foliation is the basis for classifying those which exhibit foliated structure; for the non-foliated group, the predominant mineral serves as the basis for classification.

To sum up, the outcome for the student should be that, in the broadest sense, separate cemented grains mean sedimentary rocks; foliation means metamorphic rocks; fresh, sharp-cornered grains in fine background mean volcanic rocks; and intergrown silicate crystals mean igneous or metamorphic rocks.

Making artificial rocks

4.24 *Igneous rocks*

A quick demonstration of the crystallization of alum solutions is somewhat analogous to the formation of coarse-grained and fine-grained igneous rocks. First, fill a large test-tube one-fourth full of powdered alum and cover the alum with boiling water. Hold the test-tube in a flame so that the mixture will boil gently. Slowly add just enough boiling water to dissolve the alum. Pour half of the solution into a shallow cup. Place a piece of string partly in the liquid. Stir the alum solution in the cup so that it will cool quickly, or place it in the refrigerator for more rapid cooling. Hang another piece of string in the test-tube so that it reaches the bottom. Place the test-tube where it will cool slowly. Examine the results the following day. If no results can be observed, allow to stand longer. (Better results will probably be obtained if a 'seed' crystal is placed in the solution.)

Classification of metamorphic rocks

Foliated (banded or platy):

1. Coarsely banded (bands irregular in thickness)	Gneiss
2. Schistose (regular banding, medium in thickness, and platy)	Schist
3. Slaty (regular fine banding and platy)	Slate

Non-foliated (massive or granular):

1. Chiefly calcite or dolomite	Marble
2. Chiefly quartz	Quartzite
3. Chiefly serpentine and/or talc	Serpentine or talc
4. Chiefly organic (grey or black)	Graphite and anthracite coal

---

In relation to this study of igneous rocks, it would be advantageous to have several pupils try crystal growing, e.g., sodium chloride crystals, sugar crystals, etc. (see also experiments 2.45-2.51).

4.25 *Sedimentary rocks*
This experiment may be carried out in several ways.

A. Obtain a series of sedimentary rocks of varying coloration (the colour gives a more dramatic effect). Grind these rocks up, keeping the different colours separated. (The rocks may be ground by rubbing them against a harder rock or by pulverizing them with a hammer—this is a good example of mechanical breakdown.) Collect the powdered material, and place the various coloured powdered particles in a glass jar in any desired sequence. Add water slowly, allowing it to trickle down the side so as not to disturb the layering; fill the jar slowly until the sediments are soaked, allowing the water to rise about 1 cm above the sediments. Place in the sun or near a source of heat, and allow the water to evaporate. When all the water has evaporated, break the jar; to do this safely, tie a cloth or paper bag around it and strike it in several places with a hammer.

This experiment could be repeated by the same individual, or the class could be divided into groups using the same technique. Try putting salt in the water (point out that now a solid is dissolving in solution and that this is an example of chemical breakdown of solids into solution) and allow the salt to act as the cementing agent.

B. Obtain a small bag of Portland cement. Let the pupils mix it with water and put it in tin can covers, paper cups, or small pasteboard boxes until it hardens. Study its appearance and its properties. Break a piece of cement and study it. Mix some dry cement with about twice as much sand or gravel; this will form concrete. After adding water and mixing thoroughly, place it in moulds. Allow these to harden for several days. Again study the appearance and the characteristics of the samples. (See experiment 2.66.)

C. Secure some plaster of Paris and mix it with a small amount of water. The plaster must be worked rapidly or it will harden while being mixed. Place the mixture in moulds and let it set until very hard. Study the appearance and properties of the samples. (See experiment 2.66.)

4.26 *Metamorphic rocks*
A shaped piece of clay which has first been dried can be fired quite satisfactorily by setting it on small pieces of broken pottery and heating it in a large crucible or flower pot over a bunsen burner. A kiln, if available, will be much better.

Things to do

4.27 *Making a rock collection*
A collection of the common rocks found in the locality can be made by asking each pupil to bring in one piece of rock. Explain to the pupils that it will not be necessary to know the names of all the rocks. Similar specimens may be placed together on a table. Divide the collected rocks into different groups based on shape, colour and other characteristics. Try to find as many ways as possible of grouping the rocks.

4.28 *Making individual rock collections*
Pupils should be encouraged to make their own collections of rocks. Small pasteboard or cigar boxes will serve to keep the collections. The specimens may be kept separate by putting partitions in the boxes. As a pupil identifies the rocks in his collection, he should cut small pieces of paper or adhesive tape and fasten one to each rock. Place a number on each and then paste a list on the cover of the box. The collections should be kept small. Pupils may be encouraged to exchange samples with others in order to complete their collections.

4.29 *Studying a single rock*
Select a single rock and try to learn as much as possible about it from careful observation. If it is flat, it is probably a piece or layer from some sedimentary formation. Such rocks were formed by

the hardening of sediments laid down millions of years ago. If the rock appears to be made of fine sand grains cemented together it is probably sandstone. If it is made up of larger pebbles cemented together, it is probably another sedimentary rock called conglomerate. If the rock appears to be rounded, it is probably the result of the stream action of water. Examine the rock with a magnifying glass. If it contains little flecks and crystals, it is a granite-like rock and was probably pushed up from deep in the earth long ago. Careful observation of several rocks in this manner will interest pupils in their further collection and study.

4.30 *Examining sand with a magnifying glass*
Examine a small amount of sand with a magnifying glass or under the low power of a microscope if there is one available. The nearly colourless crystals are those of a mineral called quartz which is the commonest mineral on earth. Crystals of other minerals can often be found in sand. See if you can find any others.

4.31 *The test for limestone*
You can test the rock samples to see if any are limestone by dropping lemon juice, vinegar, or some other dilute acid on them. If any are limestone they will effervesce or bubble where the acid is placed on them. The bubbling is caused by carbon dioxide gas which is given off by limestone when in contact with acid. Marble, a metamorphic rock made from limestone, will also respond to this test.

4.32 *Sorting sediments*
The sorting of sediments in the formation of sedimentary rocks can be shown by thoroughly mixing equal portions of gravel, coarse sand particles, and clay particles. Place this mixture in a glass jar (not more than half full). Fill the jar with water. Place a cap on the jar and shake vigorously. Allow the material to settle. The components will arrange themselves in order, with the heavier particles at the bottom and the clay particles on the top.

4.33 *Piezoelectricity*
An interesting experiment which pupils enjoy doing is related to the piezoelectric or pyroelectric phenomena shown by some minerals, notably tourmaline and quartz. Temperature or pressure changes cause such minerals to acquire an electric charge (positive and negative poles) when they are warmed or cooled or pressed. This may be demonstrated by dusting the cooling or warming crystal with a dust of red lead and sulphur which has passed through a silk or nylon screen.

A simple bellows can be made from a plastic nasal spray or deodorant bottle in which the aperture has been enlarged to allow a sizeable spray to be emitted. Place in the bottle a mixture of about 2 parts red lead to 1 part sulphur, place a small piece of silk or nylon stocking over the mouth of the bottle, and tighten this with a rubber band The dust particles receive electric charges as they pass through the screen formed by the stocking, and they settle on the end of the crystal which attracts them: the red lead gets a positive charge and goes to the negative end of the crystal, and the sulphur gets a negative charge and settles on the positive end of the crystal.

The demonstration is easy to perform and is highly spectacular. Spray the mineral before and after the temperature or pressure has been applied and discuss the phenomenon observed with the class.

4.34 *What are fossils and how are they formed?*
A fossil is any evidence of a form of life that lived some time in the past. Most fossils are found in layers of sedimentary rock. Fossils which are formed by burial are usually found when the sedimentary rock containing them is split open.

Cover a leaf with vaseline and place it on a pane of glass or other smooth surface. Make a circular mould about 2 cm deep and place it around the leaf. Hold the mould in place by pressing modelling clay around the outside. Now mix up some plaster of Paris and pour it over the leaf. When the plaster has hardened, you can remove the leaf, and you will have an excellent leaf print. Some fossils were made this way by having silt deposited over them, which later hardened into sedimentary rock.

Repeat this experiment using a greased clam or oyster shell to make the imprint.

If you live in a locality where fossils are plentiful, it will be interesting to have the pupils make a fossil collection for the school museum.

4.35 *Where to find fossils*
In some localities, fossils may be found in stone quarries or where there are rock outcrops. Try to find someone in the community who knows about fossils and then plan a field trip with the class to collect some of them. If there are no fossils in your locality, you may have to depend on state or national museums to send you a few. A letter to the state or national museum may prove helpful.

Soils

4.36 *Types of soils*
Obtain samples of soil from as many places as possible and put them in glass jars. Try to get samples of soil from swamps, hillsides, woods, meadows, dunes, river flats, and other localities. In this way you will gather examples of sandy, loamy, and clay soils, as well as soils rich in decayed matter or humus. Let the pupils study the samples and examine particles from each sample with a magnifying glass.

4.37 *How soil is formed from rocks by heating*
Make some rocks very hot in a fire and then pour cold water on them. The rocks will often break up both when being heated and when being cooled. One of the stages in the formation of soil is the breaking up of rocks under differences of temperature.

4.38 *How soil is formed from rocks by mechanical action*
Find some soft rocks such as shale or weathered limestone in your locality. Bring them into the classroom and have the pupils crush and grind them up into small particles. Let the pupils investigate ways in which rocks are broken up in nature.

4.39 *The effect of soil on growing things*
Obtain samples of soils from a flower or vegetable garden, from a wood, from a place where foundations are being dug, from a sandy place, from a clay bank, etc. Place the samples in separate flower pots or glass jars. Plant seeds in each type of soil and give each the same amount of water. Note the type of soil in which the seeds sprout first. After the plants have started to grow, note the soil sample in which they grow best. Record rates of growth of plants in different soils.

4.40 *Nutrition from the soil*
The rate of plant growth reflects the ability of plants to extract nutrients from rocks. Grind up specimens of the following rocks: quartzite, schist, basalt, limestone, and place the four samples separately in small glasses. Plant radish seeds in each glass, water as needed, and note rates of plant growth. Pupils should find out the chemical composition of the rocks, either by looking in a book or by testing, and explain the differences in the rate of plant growth.

4.41 *The variability of soil particles*
Obtain a glass jar that holds about 2 litres. Place several handfuls of soil in the jar. Fill the jar with water, and then thoroughly shake up the soil in the water. Let the jar stand for several hours. The size, sphericity and density of soil particles determines the order in which they will settle. The largest, more angular, and densest particles will settle out first, and will be on the bottom. The layers in the jar after settling will show decreasing size, angularity, and density from bottom to top. Examine a small sample from each of the layers with a magnifying glass.

4.42 *Soil changes with depth*
A good soil auger may be constructed from a carpenter's wood drill bit. This should be welded to a steel shank about 2 cm in diameter and about 50 cm in length. A cross-member welded to the shank will provide leverage to rotate the auger when it is drilled into the ground (see figure).

By simply turning the auger down into the earth and pulling it out of the ground at intervals,

A soil auger

soil samples which adhere to the bit can be extracted from various depths. A grid for a specific area of the earth can be made, and soil depth samples taken and compared to give a picture of subsurface conditions for that area. Afterwards, these individual soil samples can be mounted as models, or the simple observations can be recorded.

4.43 Do soils contain air?
Place some soil in a glass jar or bottle and slowly pour water over it. Observe the air bubbles that rise through the water from the soil.

4.44 Fertility varies from subsoil to topsoil
Obtain a sample of good topsoil from a flower or market garden. Obtain another sample of soil from a depth of about 50 cm. Place the samples in separate flower pots and plant seeds in each. Keep the amount of water, the temperature, and the light equal for each sample. See which soil produces the healthier plants.

Soil and water

4.45 Soil may contain water
Place some soil in a thin glass dish and heat it slowly over a small flame. Place an inverted glass jar over the dish, and water will be seen to condense on the cool sides.

4.46 Comparing the absorption of soil samples
Collect various samples of soil from different areas. Use cans as containers, and weigh each one

before collecting. Place equal amounts of soil in the containers. Heat the cans and soil in an oven at a temperature of 105-120° C until all the soil is dry. Compare the weights of each sample before and after. Compare one sample with another. Compare soil samples sheltered from rainfall with unsheltered soil samples. Correlate the absorption of these samples with daily rain gauge data. For example, how does a 25 mm rainfall affect the absorption of exposed soil, contrasted with an unexposed control sample?

4.47 Soils—acid or base?
Take soil samples from various locations. Place the samples in small jars, one tablespoonful per jar. Add an equal amount of water to each jar, sufficient to cover the material. Let the samples stand for a few minutes. Shake the jars thoroughly then drain off the liquid. The soil samples may be filtered as well as allowed to settle. Test the collected liquid with litmus paper. Blue litmus paper turns red when dipped in acidic solutions. Red litmus paper turns blue when dipped in basic solution. A neutral solution will have no reaction on either red or blue litmus paper. (See experiment 2.44.)

4.48 Water rises to different heights in different types of soil
Place about 15 cm depths of different types of soil in a series of lamp chimneys, after tying a piece of cloth over the bottom end of each chimney. Soil samples such as sand, loam, gravel (fine), clay, etc. may be used. Next stand the lamp chimneys in a pan which contains about 3 cm depth of water. Note the type of soil in which water rises highest due to capillarity. Clear plastic drinking straws can also be used for this experiment.

4.49 Which types of soil hold water best?
Tie cloth over the bottom end of several lamp chimneys, and then fill each one to within 8 cm of the top with different types of soil. Use sand, clay, loam, and soil from the woods. Under each chimney, place a dish to hold surplus water. Next, pour measured amounts of water into each chimney until the water begins to run through.

Observe the soil type into which most water can be poured without running through.

4.50 *Running water changes the soil*

A. After a heavy rainfall, let pupils take samples of running muddy water in glass jars. Let the water stand for several hours until the sediment has settled and may be observed by the pupils.

B. Construct the two trays shown in the drawings. Insert putty in the cracks to make them water-tight. Pails or glass jars with funnels may be used to collect the run-off water.

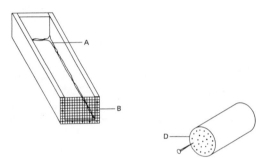

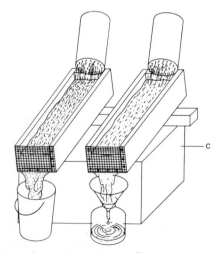

Trays for erosion and run-off tests
A putty the corners
B tack screen here
C box or crate
D watering can

1. Fill one tray with loose soil and the other with firmly packed soil. Tilt both trays slightly and using a sprinkling can pour the same amount of water on each. Observe which soil is carried away faster and the nature of the run-off water.
2. Fill both trays with soil, but cover one with sod. Water equally as before and observe both the erosion and the run-off water.
3. Fill both trays with soil but give one more slope than the other. Water and observe as before.

4.51 *Raindrops may affect soils differently*

Fasten a sheet of white paper to a piece of stiff cardboard with paper clips. Lay it flat on the floor. Drop coloured water on it with a medicine dropper. Note the size and shape of the splashes. Repeat, but this time prop up one end of the cardboard. Study the effect on the splashes of varying the height from which the water is dropped, and also varying the slope and the size of the drops.

Try different combinations of the variables. A record of the results may be kept if a clean sheet of paper and different coloured water is used for each situation.

4.52 *Splash sticks*

Get the children to paint several metre-sticks white (so that mud splashes will show plainly), and place the white sticks outdoors in various areas. Hold each stick upright by means of brick and rubber band, as shown in the illustration. After a rainstorm, tell the children to note the height to which mud has been splashed on each

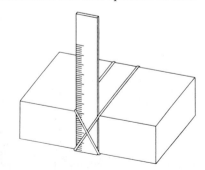

The splash stick must be held upright

stick. Make a chart showing the height to which mud was splashed on different sticks in a grassy area, a sandy area, a garden area, and others that you might think of.

Do all rainstorms produce the same effect in the same place? The children should repeat the activity during successive rainstorms, or use different forces of water from a hose.

4.53 *Erosion and conservation*
A. Take a field trip to a locality where running water has caused damage by cutting gulleys. Pupils should note the damage and think of ways to prevent it. Why did it happen? How could it have been prevented? What can still be done?

B. In the laboratory, pupils should use equipment to design ways of preventing erosion. Specifically, contour ploughing, terracing, and crop rotation should be considered. Some of the designs may be as follows.
1. Fill the trays described in experiment 4.50B with loose soil, and tilt both to the same angle. Make furrows with a stick, running up and down the hill in one box, and across the hill in the other box. Sprinkle each with the same amount of water. Observe the erosion in each case, and the runoff water
2. Again fill the trays with loose soil. Water them until well-defined gulleys are formed by the running water. Now block the gulleys at intervals with small stones and twigs. Again water, and observe the effect of blocking the gulleys.

C. In almost every school yard, there is a place where running water has caused damage. Enlist the class in a project to decide upon means for preventing the erosion, and then let them carry out their project.

4.54 *Soil permeability*
Obtain three cans of equal size and cut out the tops and bottoms. Across the bottom section of each can, fasten some fine screen material by wrapping a wire around the lower rim. To prevent fine particles of soil from passing through the screen, place a piece of filter paper inside the can on top of the wire mesh.

Collect three samples of soil—coarse, medium, and fine. Heat these in an oven at a temperature of 105-120° C until the samples are thoroughly dry. Place equal amounts of the three soils in different cans. Mount all three samples so that water can be poured into them and collected underneath. Pour equal amounts of water into each container. Record, in each instance, the time it takes for the water to stop filtering through. Compare the amounts of water collected under the cans.

4.55 *Minerals in solution*
Minerals in solution are best demonstrated by computation. Pupils can find out from the local water supply service the total amount of dissolved mineral matter per unit volume in the untreated drinking water; let them compute the weight of mineral matter in the total amount of water consumed by the community in one year.

The solution of limestone by weak acids can be demonstrated by crushing a small piece of limestone, placing it in a small beaker, and covering well with weak hydrochloric acid (4 parts water to 1 part concentrated acid); allow to stand until the rock is dissolved. Note the residue of insoluble matter, mostly clay and quartz.

4.56 *Capillary action, solution, and deposition by ground water*
Place a mixture of table salt and fine, dry sand in the bottom of a small aquarium to a depth of 2 to 5 cm. Cover this layer with about 5 cm of clean sand (no salt). Insert a glass tube with a funnel, supported by a clamp stand, into the sand at one corner of the aquarium (see overleaf). Make sure that the tube reaches the salt layer. Clamp a heat lamp on a stand so that it can shine down on the other side of the aquarium.

Pour water into the funnel (the tube may have to be shaken slightly to get the water to move down the tube). Observe the side of the aquarium; the water can be seen to move through

4.56   Capillary action, solution and deposition

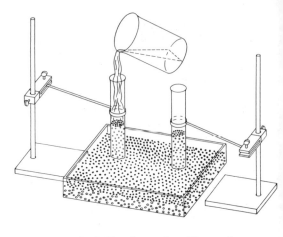

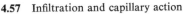

4.57   Infiltration and capillary action

the sand. Put in sufficient water to wet a layer about 2 cm deep along the bottom of the aquarium.

Light the lamp and let it burn for several hours. In the vicinity of the lamp, the water will rise through the sand by capillary action, bringing the salt in solution up with it. The heat will cause the water to evaporate, and the salt will be deposited near and at the surface. Taste some of the sand near the heat lamp to see if it is salty. In nature, the sun has the same effect as the heat lamp in this experiment.

4.57 *Infiltration and capillary action by ground water*

Fill two glass tubes (approximately 2 cm in diameter and 30 cm long) about half full of dry, fine sand. Support the tubes vertically by clamp stands, with their bottoms resting in some type of flat dish or aquarium (see diagram). Pour water into one tube. The water will infiltrate down through the pore spaces of the sand, move into the dish, and partially move up the other tube by capillary action.

4.58 *Oxidation*

Oxidation can be demonstrated by placing a small piece of steel wool in a small box of clean sand and moistening it daily. Note the effect upon the steel wool and the staining of the sand. Let pupils powder a small piece of pyrite, place it on a watch glass, and moisten daily. After a period of several weeks, note the development of a white crystalline substance, iron sulphate. Oxidation of iron minerals is generally accompanied by a change of colour to yellow, brown, and red oxides and hydrated oxides. (See experiments 2.42 and 2.318.)

4.59 *An effect of freezing water*

Fill a bottle with water and put the top on securely. Wrap the bottle in a cloth (to prevent the shattered glass from falling). Put the bottle into the freezing compartment of a refrigerator. After 24 hours, remove the bottle and examine it. What can be observed? Why is the bottle cracked? What caused the pressure? Why did the ice exert a force? (See experiment 2.129).

Additional activities

4.60 *Mud cracks*

Use collection trays to collect several samples of various types of clay and fine silt. Fill the trays about three-quarters full with the collected materials. Add sufficient water to cover the material.

Place the collection trays in direct sunlight. Observe how the mud cracks form. Compare the number of formed mud cracks from tray to tray. What angles are formed as the mud cracks appear? Are they all the same?

4.61 *Soil horizons*

Mature soils usually exhibit a well-marked profile consisting of three main layers or horizons designated A, B, and C. These differ in colour, texture, and structure and vary in thickness.

The A horizon is called the topsoil. Soluble materials are removed. Generally, topsoil is rich in organic matter and in soil organisms. The B horizon is called the subsoil. This horizon accumulates clay washed out of the top soil above. Iron minerals are usually present and most likely they will be oxidized. The C horizon is the unconsolidated, weathered parent material.

Make models of soil horizons from various places and compare the depths of the A and B horizons. This is best accomplished by observing the horizons in fresh road cuts or in gulleys. Using a spade, make a clean vertical cut to expose the various horizons. Let the material dry. Obtain a board or other rigid base and smear permanent glue on one side of it. Press this side against the cut so that numerous particles from each horizon stick to the board. Now remove the board, and fill in any blank spots with scrapings from the cut. Let the model harden. Compare the different profiles, noting the depths of the horizons in each model as well as the composition of the material at each of them.

4.62 *The effect of plants upon erosion*

A. Pupils should observe an area where the soil has eroded because of lack of plant cover. Discuss what they see, and have them explain their ideas about why the area looks as it does. What might stop the soil from blowing or washing away? Can pupils demonstrate their ideas?

B. Let the children plant grass seed in a section of sandy soil, in an erosion table (set up as shown in experiment 4.50). Run water down the erosion table when the grass has developed a network of roots. Do the roots have any holding effect on the soil? Pull up some of the sprouting grass and have the pupils change the amount of water coming on to the erosion table. What are the effects of erosion?

4.63 *Examining life in the soil*

Soil, besides being acidic or alkaline, may have animal life of various kinds in it. The kind and amount of plant and animal life varies in different soils. The important thing is that this life is often a factor that can change the make-up of the soil.

Ask the class to examine the surface of a square metre of soil. Note any earthworm mounds, anthills, or other signs of animal activities. Carefully remove the surface plant life and examine the soil for further signs of animal life. How do earthworms cause more air to enter the soil? How do they turn over the soil? Does their ingestion of soil particles change the composition of the soil? Do their droppings and dead bodies change the composition of the soil?

4.64 *Wind deposits*

Obtain three large, shallow tins (e.g., pie tins); fill one with a litre of moist sand, the second with dry sand, and the third with flour. The pupils should then place the tins 7 metres from an electric fan which is directed to blow towards the tins. Where is the force of wind the greatest? Move each pan towards the blowing fan until there is a slight movement in the pile of material.

On a chart, indicate the three materials and the distance at which each showed movement as a result of the 'wind'. Which moves at the least distance? Which moves at the farthest distance? Why? Do the pupils notice a pattern in the way the materials have blown? The lightest is the farthest away and the heaviest is the closest. Many mixtures of particles are separated in this way. Explain that this is called sorting, and it is a frequent occurrence in nature.

The children can attempt to create sand dunes by using the fan and the dry sand. Are they able to create any patterns of ridges in the sand? What causes sand dunes in nature? If a sandy area is available, the children might enjoy studying this for evidence of the wind having formed dunes into ridges.

Astronomy and space science

Astronomy and space are always interesting topics for children studying science. In many places the basic concepts of astronomy are taught descriptively—that is, the children merely read about them. In this section, many experiments are suggested to enable the teacher to develop some of the concepts from observation and experiment.

No attempt has been made to grade the experiments. It is suggested rather that teachers select those experiments that seem most appropriate for the topics being taught.

Instruments for astronomy

4.65 *A simple refracting telescope*
To make a simple telescope, first obtain two cardboard tubes, one fitting inside the other.

It is not possible to make a satisfactory telescope unless good lenses are available, a fact which was soon discovered by early experimenters.

A linen tester (sometimes a stamp magnifier also) has lenses which are achromatic, that is corrected for colour distortion. Such a lens of focal length 2 or 3 cm will provide a suitable eye-piece when mounted in a cork with a hole in it.

For best results, it is equally important that the object glass should be achromatic. If such a lens is available with a focal length of 25 to 30 cm it should be fixed in the wider cardboard tube by means of plasticine or adhesive. A little

adjustment is required to get both lenses on the same optical axis. When this has been achieved and the focusing done by sliding the tube, you will have an instrument superior to the one with which Galileo made all his discoveries. (See also experiment 2.219.)

Jupiter's moons are readily observed with this apparatus, but not Saturn's rings.

4.66 *A simple reflecting telescope*
A simple reflecting telescope can be made from a concave mirror, e.g. a shaving mirror. The mirror is mounted in a wooden box in such a way that it can be tilted at different angles (see diagram 4.66A). A wooden upright is attached to the box so that its angle of inclination may also be varied. Two short-focus lenses are fixed in corks which are then placed in a short length of mailing tube as an eye-piece.

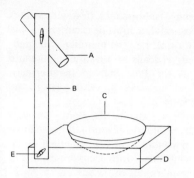

4.66A A simple reflecting telescope
  A mailing tube with lenses
  B upright
  C mirror
  D box
  E pivot

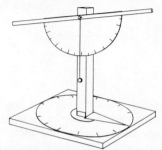

screw will show if the support used is upright, and will serve also to measure the elevation of the star or any other object observed through the drinking straw.

An improved model for finding altitude and bearing of a star can be made by fixing the upright to a baseboard with a screw and two washers, leaving it free to rotate. A piece of tin fixed to the upright will indicate the angle on a horizontal scale (see diagram). It is with rough apparatus such as this that many early discoveries were made.

4.66B Ray diagram

Attach this eye-piece to the wooden upright and make the necessary adjustments (see diagram 4.66B).

4.67 *A simple theodolite or astrolabe*
A simple theodolite or astrolabe may be made by fixing a drinking straw to the base line of a protractor with sealing wax or glue.

A plumb line hung from the head of a fixing

Sun-dials

4.68 *Demonstration sun-dials*
A. A simple demonstration of the sun-dial may be given by placing an upright stick in the ground so that it is not likely to be shaded from the sun. The position of the shadow of the top of the stick can be marked on the ground at hourly intervals (see figure). (See also experiment 4.89B).

Demonstration of the sun-dial (figure drawn for the southern hemisphere)

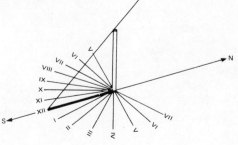

B. The simplest dial can be made from a circular metal plate divided into 24 equal arcs. A steel knitting needle is pushed through the centre of the plate so that the plane of the plate is at right angles to the needle (see figure).

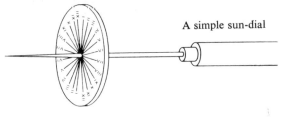

A simple sun-dial

The plate is positioned so that the gnomon (i.e. the needle) points towards the celestial pole and the noon position of the shadow of the gnomon falls on the XII marking. The shadow will then fall on the other markings at approximately the correct time.

(The plate should be marked on both sides, as the shadow of the gnomon will move from one side to the other as the sun's declination changes.)

4.69 *A sun-dial for your home*
The base should be made of a flat rectangular piece of wood, metal or polystyrene. The gnomon ABC consists of a thin triangular piece of metal or plastic laminate and such that angle ABC = latitude of the place at which the dial is being set up and angle ACB = 90° (see figure).

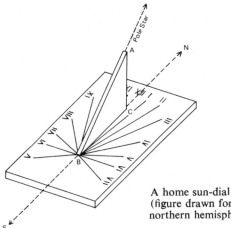

A home sun-dial
(figure drawn for the
northern hemisphere)

The base must be horizontal (test with a spirit level) and its central line must lie accurately along the north-south line, i.e., the meridian. The gnomon is erected vertically so that the hypotenuse points towards the Pole Star in the northern hemisphere and the celestial south pole in the southern hemisphere.

For approximate results, the hour markings can be made by noting the position of the shadow of the gnomon at hourly intervals, using a watch set to local mean time. More accurate results are obtained if the markings are made on 15 April, 15 June, 1 September or 24 December, when there is no difference between watch time and dial time. Errors of up to 16 minutes are possible if markings are made on other dates.

Note. If accurate hour markings are required these may be obtained by finding the angles the markings make with BC using the following formulae:

$$\tan \widehat{IBC} = \tan 15° \sin \text{lat.}$$
$$\tan \widehat{IIBC} = \tan 30° \sin \text{lat.}$$
$$\tan \widehat{IIIBC} = \tan 45° \sin \text{lat.}$$
$$\tan \widehat{IVBC} = \tan 60° \sin \text{lat.}$$
$$\tan \widehat{VBC} = \tan 75° \sin \text{lat.}$$
$$\tan \widehat{VIBC} = \tan 90° \sin \text{lat.}$$

Since the markings are symmetrical about the central line XY no other angles need be calculated. *N.B.* If the base of the dial is erected vertically then the angle between the gnomon and the base must equal (90° minus latitude of the place).

4.70 *A universal globe sun-dial*
A. With a globe of the earth you can make a sun-dial that shows the season of the year, the regions of dawn and dusk, and the hour of the day wherever the sun is shining.

The rules for setting up the globe are simple and easily followed. It is rigidly oriented as an exact model of the earth in space, with its polar axis parallel to the earth's axis, and with your own town (or state) right 'on top of the world'. First turn the globe until its axis lies in your local meridian, in the true north and south plane that may be found by observing the shadow of a vertical object at local noon, by observing the Pole Star on a clear night, or by consulting a magnetic compass (if you know the local variation of the compass). Next turn the globe on its axis until the circle of longitude through your home locality lies in the meridian just found. Finally tilt the axis around an east-west horizontal line until your home town stands at the very top of the world. If you have followed these three steps, then your meridian circle (connecting the poles of your globe) will lie vertically in the north-south plane, and a line drawn from the centre of the globe to your local zenith will pass directly through your home spot on the map (see figure).

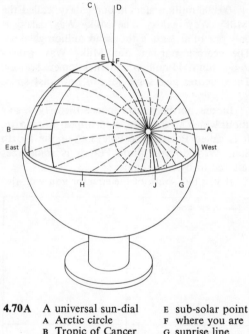

4.70A	A universal sun-dial	E	sub-solar point
	A Arctic circle	F	where you are
	B Tropic of Cancer	G	sunrise line
	C sun on your meridian	H	sunset line
	D your zenith	J	midnight sun

Now lock the globe in this position and let the rotation of the earth do the rest. This takes patience, for in your eagerness to see all that the globe can tell you, you may be tempted to turn it at a rate greater than that of the turning of the

earth. But it will take a year for the sun to tell you all it can before it begins to repeat its story.

When you look at the globe sitting in this proper orientation—'rectified' and immobile—you will of course see half of it lighted by the sun and half of it in shadow. These are the very halves of the earth in light or darkness at that moment. An hour later, the circle separating light from shadow has turned westward, its intersection with the equator having moved 15° to the west. On the side of the circle west of you, the sun is rising; on the side east of you, the sun is setting. You can 'count up the hours' along the equator between your home meridian and the sunset line and estimate closely how many hours of sunlight still remain for you that day; or you can look at a country to the west of you and see how soon the sun will rise there. As you watch the globe day after day, you will become aware of the slow turning of the circle northward or southward, depending upon the time of year (see also experiment 4.98).

B. It is not easy to appreciate the fact that the sun's rays are parallel as they fall on the earth. A simple experiment is suggested. On a bright morning, take a piece of pipe or a cardboard tube and point it at the sun so that it casts a small, ring-shaped shadow. *Warning:* Do not look at the sun through the tube as direct sun rays can destroy the retina of your eye. Now if at the very same moment someone 120° east of you—one-third of the way round the world—were to perform the same experiment, he would point his tube westward at the afternoon sun. Yet his tube and yours would necessarily be parallel to within a very small fraction of a degree. If you point the tube at the sun in the afternoon, and someone far to the west simultaneously does the same (but for him it will be in the morning), his tube will again be automatically parallel to yours. This experiment will help explain how it is that, when our globes are properly set up, people all over the world who are in sunlight will see them illuminated in just the same way.

C. It is easy to tell from the global sun-dial just

how many hours of sunlight any latitude (including your own) will enjoy on any particular day. All you need to do is to count the number of 15° longitudinal divisions that lie within the lighted circle at the desired latitude. Thus at 40° north latitude in summer the circle may cover 225° of longitude along the 40th parallel, representing 15 divisions or 15 hours of sunlight. But in winter the circle may cover only 135°, representing nine divisions or nine hours. As soon as the lighted circle passes beyond either pole, that pole has 24 hours of sunlight a day, and the opposite pole is in darkness.

Becoming familiar with the stars and planets

4.71 *Recognizing the main constellations and making a star map*

This is a convenient home task, and is best done about the time of the new moon. The moonlight does not then interfere with a good view of the stars. It is helpful to take outside a piece of brown paper with pinholes pricked through it in the form of a few of the constellations. When the paper is held up to any light the pinholes become visible, and the paper can be rotated until a similar star pattern is recognized. This is particularly

simple for the northern hemisphere where the Pole Star will be in the centre very close to the North Celestial Pole (see diagram below, left).

In the southern hemisphere the easiest way is to start with the Southern Cross, which consists of four stars. Three of them are very bright. They are shown below, right, which also indicates how to locate the South Celestial Pole in a rough manner (see also experiment 4.78).

After a few constellations have been learnt in this way, it is instructive to make one map in the early evening, and one just before going to bed.

Our planet, earth, rotates on its axis from west to east as it revolves about the sun. The sun is our own personal star. Our sun is just one of 100,000 million stars in our galaxy, called the 'Milky Way' galaxy. The Milky Way galaxy is only one of at least a thousand million galaxies. The nearest star in our Milky Way galaxy to our sun is 43 million million kilometres away. These figures give some idea of how vast space is.

In the northern hemisphere there is a star about which all the other stars appear to revolve. It is called Polaris, the Pole Star or the North Star. Why do you think Polaris is sometimes referred to by these names?

If you live south of the equator you will also

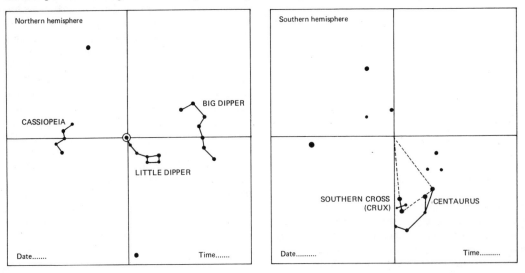

find that the stars seem to revolve about a point in the sky, although there doesn't happen to be a star at that particular point.

The stars appear to make one full revolution every 24 hours and likewise once every year. This accounts for the fact that the various constellations (groups of stars) are not seen in the same position at different times of the night and at different times of the year. (An explanation of how to fix the position of a star on the celestial sphere is given in experiment 4.74.)

4.72 *Locating some constellations from north of the equator*

The Pole Star is really the key to locating and identifying constellations and individual stars for those who live north of the equator. The following notes will help us to recognize a few of the constellations. The most obvious constellation is Ursa Major, commonly called the 'Big Dipper' or the 'Plough'. The Big Dipper is a constellation finder. It is very valuable in locating the Pole Star.

After locating the Big Dipper, note the two stars that form the front edge of the dipper cup. If you follow these in a straight line, they will lead you to the North or Pole Star (Polaris). Once we have located the Pole Star we can easily locate other constellations.

Actually there are two dippers in the sky. These are referred to as the 'Two Bears', because ancient observers thought that the outlines of the constellations resembled bears. We have a Great Bear (Ursa Major, or the Big Dipper) and a Little Bear (Ursa Minor, or the Little Dipper).

Having located the North or Pole Star from

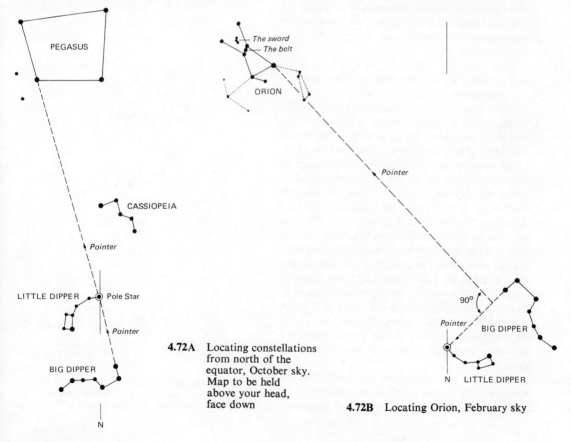

4.72A Locating constellations from north of the equator, October sky. Map to be held above your head, face down

4.72B Locating Orion, February sky

the Big Dipper we can always find the Little Dipper because the Pole Star is the last star in the handle of the Little Dipper. The Little Dipper always appears to pour into the Big Dipper.

We next look for Pegasus, the mythological winged horse. In diagram 4.72A, which refers to October, you can see that the four stars of Pegasus form a box. The north-east star also belongs to Andromeda. Pegasus is found by continuing the straight line from the two stars which form the outer edge of the Big Dipper cup through and beyond the Pole Star.

Let us now look for the constellation Cassiopeia. This is easily found, for it is located opposite the Big Dipper beyond the Pole Star. It forms the letter W and is often called 'Cassiopeia's chair'.

Another familiar and easily recognizable constellation is Orion, the 'Great Hunter'. Orion contains three bright stars in a line called the 'Belt of Orion'. Below the 'belt' are three fainter stars, called the 'Sword'. (See diagram 4.72B on the preceding page.)

4.73 Locating some constellations from south of the equator

The obvious key constellation is the Southern Cross (Crux). In the beginning of December you will see it low down on the southern horizon at midnight. When you have found the Southern Cross you can easily locate two bright stars of Centaurus. They are almost in line with the two lower stars of the Southern Cross towards the south-west. The two stars of Centaurus are also called the Pointers of the Southern Cross. The one which is farthest away from the Southern Cross is very near to the earth by astronomic standards. For a long time it was thought to be our nearest star. It takes the light more than four years to travel from this star to the earth although light travels at the tremendous speed of 300,000 km/s. Astronomers measure a large distance by giving the time light takes to cover it. In this case they say that the star is more than four light years away.

From the Southern Cross we can follow the Milky Way to the north and find Canis Major or the Great Dog. This constellation is of particular

interest since it contains Sirius, also called the Dog Star. It is the most brilliant star there is. Only a few stars are nearer to us than Sirius, which is 8.5 light years away.

Not far from Canis Major you will find Orion which can also be seen north of the equator.

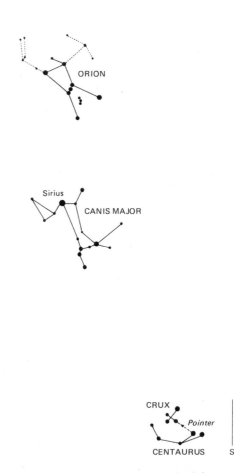

4.73 Locating constellations from south of the equator, December sky. Map to be held above your head, face down.

4.74 *Locating some constellations within the tropics (and sub-solar positions)*

For purposes of identification the stars can be imagined as being situated on the inside of a sphere, called the celestial sphere, which is concentric with the earth. The Pole Star, which is positioned approximately at the north pole of the celestial sphere, is almost directly above the north pole of the earth, and the celestial equator circles the celestial sphere directly over the earth's equator.

A point on the earth's surface can be identified by its latitude and longitude. The longitude is sometimes called the meridian and is the line joining the north and south poles and passing through the point; e.g., the 0° longitude, or Greenwich meridian, passes through the north pole, a place called Greenwich in England, and the south pole. In the same way a star's position on the celestial sphere is determined by its declination (which corresponds to latitude and is measured north and south of the celestial equator) and its right ascension (which corresponds to longitude).

A point on the celestial sphere immediately overhead to an observer on the earth is called the zenith of the observer. Thus the Pole Star would be at the zenith of an observer at the north pole of the earth, and at approximately midday on 15 May the sun would be at the zenith of an observer in a place of latitude 20° N.

The star chart which will be found inside the back cover of this book represents that part of the celestial sphere which is seen by an observer on the earth's equator. It extends from 35° N. to 30° S. and thus does not have the distortion of the constellations within the corresponding declinations which is usually found in star maps for northern and southern latitudes. This makes the chart of special interest for those living in the tropics where weather conditions often limit observations to within 45° of the zenith. Orion's belt, when visible, gives an approximate east-west direction and the line joining the midpoints of the shorter sides of the Orion quadrilateral gives a useful guide to the north-south direction.

The distances are measured in angular degrees and the equator is divided approximately into months. Each date sets the chart at midnight for an observer on the equator, i.e., whose zenith is on the equator. For other places the zenith at midnight is given by the intersection of the parallel of latitude of the observer and the meridian cutting the equator at a particular date, e.g., Rigel is at the zenith at midnight on 7 December for places situated along the latitude 8° S.

The stars visible on the observer's meridian at 11 p.m. on any night will be those seen on the meridian at midnight 15 days earlier. Thus, Betelgeuse which is approximately at the zenith at midnight on 17 December will be in the same position at 11 p.m. 15 days later, i.e., on 1 January.

The bold dashed curve enables an observer to determine approximately the latitude at which the sun will be directly overhead at midday on any day. This is determined by noting the intersection of the curve with the latitude parallels, e.g., the sun is overhead at midday at places on latitude 20° S. on 25 January.

Note. The curve for determining the position of the sun each day must not be confused with the ecliptic with which it is symmetrical. This explains the apparent twelve-hour error in the right ascensions of the stars as plotted.

4.75 *The apparent daily rotation of the sky*

The materials required are a star chart, a plumb line (string and weight), pencil and paper and a watch or clock.

A. Choose a place where you have a clear view of the northern sky (or the southern sky if you are south of the equator), including parts close to the horizon. Locate as accurately as you can your celestial pole and rig up the plumb line so that it appears to go through the pole (Pole Star, if north of the equator) seen from your view point. Observe carefully where the lower end of the plumb line appears against the stars. Draw a line on the star chart to represent this position of the plumb line, and note the time to the nearest minute. Make the same type of observation two or three hours later, mark a line on the chart and note the time. Also record the calendar date. In which direction does the sky appear to turn (clockwise or counterclockwise)? Relate this to the earth's rotation: if you looked down on the north (or south) pole of the earth, how would the earth seem to turn (clockwise or counterclockwise)?

Measure the angle between the two lines with a protractor. How many degrees is it? Calculate the change in degrees per hour. From this, calculate the time required for one complete rotation (360°). What conclusion can be drawn from this result? How accurate do you think your result is? This observation may be supplemented by photographic star trails (see also experiment 4.90).

B. Stand in a particular spot that can be exactly identified later. If you are in the northern hemisphere, identify one of the prominent constellations in the southern sky (e.g., Orion) and sketch its position relative to prominent landmarks (buildings, trees, etc.). In the southern hemisphere identify a constellation in the northern sky. Note the time. Two or three hours later, standing at exactly the same spot, again observe the same constellation, sketch its position, and note the time. How does the change of position of this constellation fit in with the change you observed in A above? Would its period for one complete otation differ from that found above?

C. Observe a constellation that appears low in the eastern sky, and observe it two hours later. Explain the changes.

D. Observe a constellation that is about midway up in the western sky, and observe again two hours later. Explain the changes.

4.76 *Making a constellarium*
A constellarium is a simple device used in teaching the shapes of various constellations.

A. Obtain a cardboard or wooden box and remove one end. Draw the shapes of various constellations on pieces of dark-coloured cardboard large enough to cover the end of the box. Punch holes on the diagrams where the stars are located in the constellations. Place an electric lamp inside the box. When the lamp is turned on and various cards are placed over the end of the box, the constellations may be seen clearly.

Another way is to obtain several tin cans into which an electric lamp may be fitted. Holes are punched in the bottoms of the cans to represent the stars in various constellations. When the lamp is placed inside a can and switched on, the light shows through the openings and the shape of the

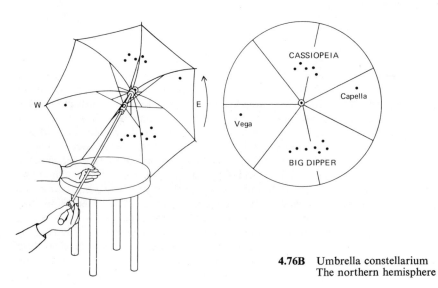

4.76B   Umbrella constellarium
The northern hemisphere

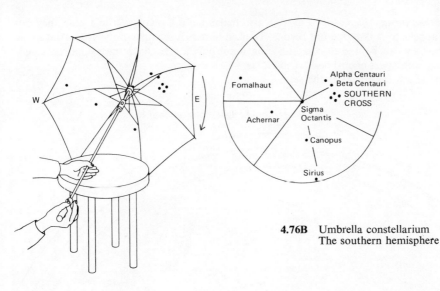

4.76B   Umbrella constellarium
The southern hemisphere

constellations may be observed. The cans may be painted to prevent rusting and kept from year to year.

B. Since an umbrella has the shape of the inside of a sphere, it can be made into a constellarium that will illustrate portions of the heavens and how they move. You will need an old umbrella that is large enough for this purpose.

The northern hemisphere. Using chalk, mark the North Star, or Polaris, next to the centre on the inside of the umbrella. Consult a star map, and mark the star positions for various constellations with crosses. When you have filled in all the polar constellations, you can paste white stars made from gummed labels over the crosses, or you may paint the stars in with white paint. Later you can draw dotted lines with white paint or chalk to join the stars in a given constellation.

If the handle of the umbrella is rotated in a counterclockwise direction, you will see how the various stars trace a circular path about the Pole Star.

The southern hemisphere. South of the equator, the umbrella should be pointed towards the southern celestial pole and we will therefore have to turn it clockwise. As in the northern hemisphere,

the stars will rise in the east and set in the west. In the diagram above you can see some of the more prominent stars and constellations marked on the umbrella.

4.77 *The seasonal shift of the sky*
As the earth travels in its orbit around the sun the constellations seem to move across the sky. The materials required for observing the shift are a star chart and a plumb line.

Make observations as described in 4.75, except that you make only one set of observations and record the time.

At least one month later, repeat the same observations in the same way, at as nearly the same time of night as possible. When you compare the two observations made at the same time of night, what change do you see in one month (or more?) How much change would occur in one year, if the same rate continues? What does this mean, when you recall that we tell time by the sun? Will there be a time of year when you cannot see Orion (for example) at all? Why? Answer the same questions for the Big Dipper and North Star, if you are north of the equator. What about the Southern Cross if you are south of the equator?

4.78 *The stars tell the time and the date*

Because the stars appear to rotate one full revolution in 24 hours, they can be useful in telling time, at least during those hours of darkness when the stars are visible to us. Inasmuch as the stars also make one full revolution in a year, they can be used to tell us the time of the year. And so we have not only a star clock, but also a star calendar.

Below: star charts for the northern hemisphere and (right) the southern hemisphere

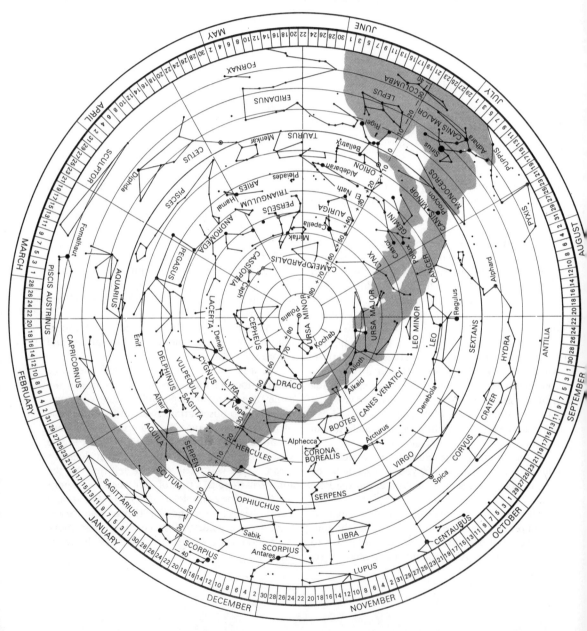

A. *The star calendar*. The dates round the edge of the star chart for the northern hemisphere show when the corresponding part of the heavens is due north at midnight. For the southern hemisphere the dates show the part which is due south at mid- night. Knowing this you can easily rotate the star map so that it corresponds to what you see in the sky. If you are north of the equator and you have to rotate the map 15° clockwise from the midnight position, the time is 1 a.m.; if you have to rotate

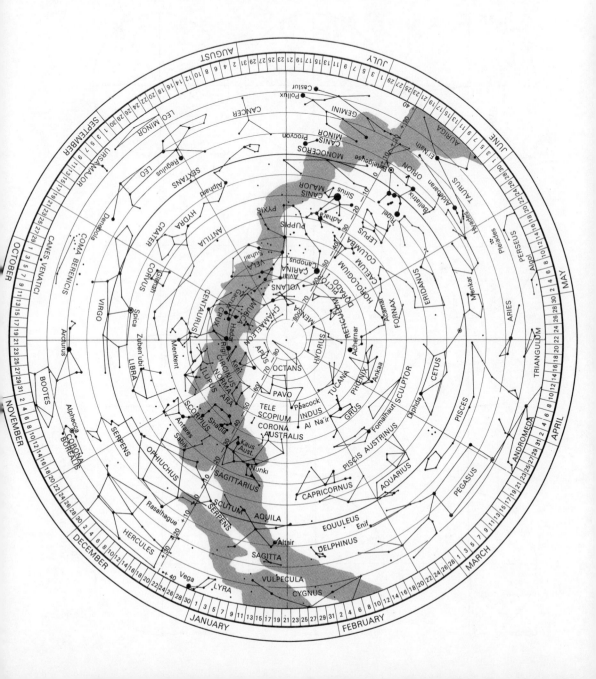

it 30° counterclockwise, the time is 10 p.m. South of the equator it is the other way round since you are facing south. If you have to turn the map 15° clockwise from the midnight position it means that the time is 11 p.m. The times determined this way are sun times and they may differ from your local standard time.

B. *The star clock*. Separate sets of diagrams are given below for the northern and southern hemi-

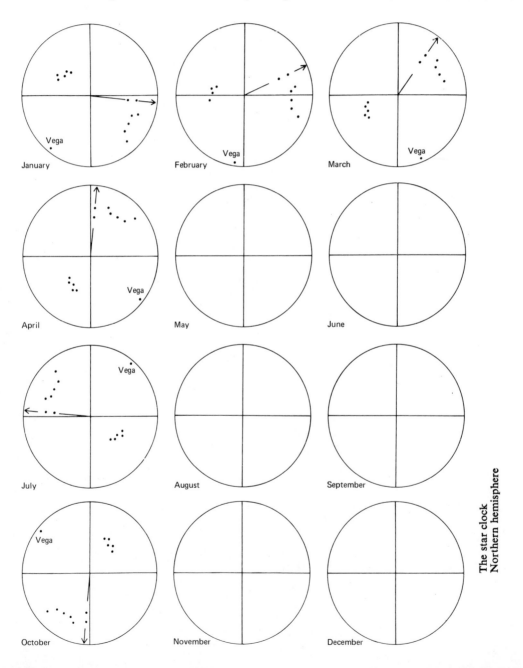

The star clock
Northern hemisphere

spheres, one clock for each month. The nine o'clock positions of the star clock's hand are marked off at the middle of some months.

Can you fill in the nine o'clock positions for May, August and November? Try to fill in mid-night positions for June, September and December. In the southern hemisphere, locate roughly the southern celestial pole (see experiment 4.71).

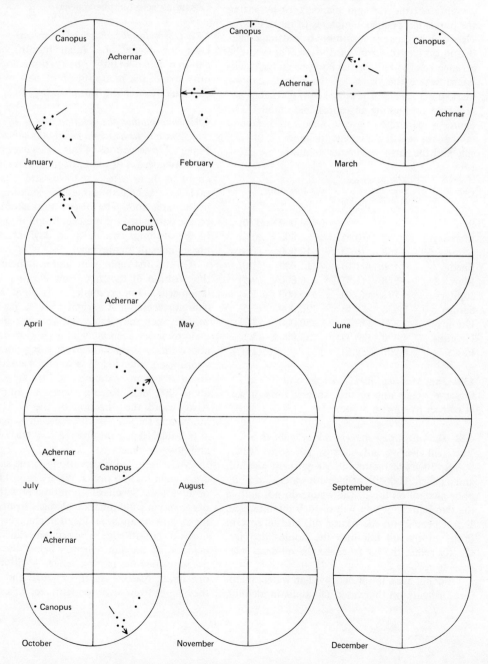

Southern hemisphere

4.79 *A model of the solar system*

The concepts of the relative size and distance of the planets from the sun can be illustrated by having pupils make a model of the solar system. This can be done: (a) by using various sized balls to represent the sun and planets; (b) by letting the pupils use perspex templates to make clay or plasticine models; or (c) simply by cutting circles of appropriate size from cardboard. These can be arranged on the wall, on the floor, or on the blackboard where the orbits of the planets can be marked with chalk. The table below gives the data necessary for making an approximate model. The figures in parentheses give a scale for distances, taking the earth's average distance from the sun and the earth's diameter as units.

Body	Average distance from the sun (in millions of km)		Diameter (km)	
Sun			1,400,000	(110)
Mercury	58	(0.4)	4,800	(0.4)
Venus	108	(0.7)	12,000	(1.0)
Earth	150	(1.0)	13,000	(1.0)
Mars	228	(1.5)	6,800	(0.5)
Jupiter	778	(5.2)	140,000	(11.2)
Saturn	1,420	(9.5)	120,000	(9.5)
Uranus	2,870	(19.2)	50,000	(3.7)
Neptune	4,490	(30.1)	53,000	(4.1)
Pluto	5,900	(39.5)		(1.0?)

4.80 *The 'Morning' and 'Evening' Star*

Observe Venus and record when it rises or sets in respect to sunrise or sunset.

4.81 *Demonstrating movements of planets*

You will need a tall, narrow jar, some water, S.A.E. 30-grade motor oil, 90 per cent alcohol, and a pencil. Half-fill the jar with water. Carefully pour alcohol on top of the water; do not agitate the two liquids or you will disturb the interface. Dip a pencil into the motor oil, and let several drops of the oil fall into the liquid-filled jar. Gently rotate the jar to cause the oil-drop 'planets' to revolve.

Alcohol has a lower density than water; therefore, it floats on the water. Oil sinks in alcohol, yet floats on water. In such a 'free' state, the oil forms spheres and stays at the interface between alcohol and water.

Observing celestial phenomena

4.82 *Observing the phases of the moon*

Over the period of a lunar month, have the children make nightly observations and sketch drawings of the moon. Begin at new moon and continue through the four phases.

4.83 *Determining the relationship of the moon's phase to its apparent position in the sky*

Make all observations of this series over an interval of two weeks or more. Begin the observations about an hour after sundown, and observe on each clear night at the same time, standing always in the same spot. The observation should begin on the date when the crescent moon is just visible in the evening, two or three days after a new phase (consult an almanac).

On the first night, note and sketch accurately the position of the moon relative to prominent landmarks. (For example, it is right above a church steeple, or it is right above a point midway between the steeple and an office building.) Determine as accurately as you can its height above the horizon in degrees, using your fist or your fingers extended. (Fist at arm's length equals about 10°, span of thumb and little finger equals about 20°, etc.) Record these on your sketch. Also record the direction of the horns of the moon, and the shape of the crescent as accurately as possible. Repeat the observation two hours later and note the time.

Make repeated observations in the same way every night for two weeks. Write a report on these observations. Specifically point out how the shape of the moon's illumination changes from night to night, how its apparent location changes, how its horns or cut-off edge are oriented relative to the position of the sun which is below the western horizon, how the moon changes position during one night, reasons for this change and also for the change from night to night, etc. At some time about the last quarter-phase of the moon (see the almanac) observe in the early morning in the same way as before. How do these observations fit in with those made in the evenings?

4.84 *Observing a solar eclipse*
Explain to the children that by observing eclipses, the time intervals in which they occur, and the shadows they cause, scientists have been able to gather some information about the shape, size, and motions of the sun, moon, and earth (see figure).

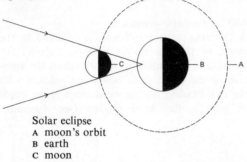

Solar eclipse
A moon's orbit
B earth
C moon

Ask the children if they can tell one way in which the shape of the earth has been determined. Ask them to watch in newspapers or look in an almanac for the dates of eclipses. Plan to take the class outdoors when an eclipse occurs in your area.

Warning. Do not allow pupils to look directly at the eclipse, as damage may occur to their eyes. Even the use of smoked glass or several layers of exposed film over the eyes is not entirely safe.

One safe method of observing an eclipse is to view it indirectly. Have the children punch a hole through a piece of cardboard. Then ask them to turn their backs on the sun and hold the cardboard over one shoulder to permit the sun's image to shine through the hole on to a second piece of cardboard or paper held in front of them. Do not allow them to look at the sun through the hole in the cardboard (see also experiment 4.96).

4.85 *Observing a lunar eclipse*
In this case direct observation is safe. Get the children to note the shape of the earth's shadow

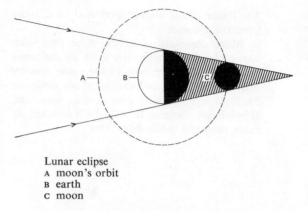

Lunar eclipse
A moon's orbit
B earth
C moon

as its edge crosses the moon—evidence that the earth is round, though the effect could be caused by a disc-shaped earth (see figure). (See also experiment 4.95.)

4.86 *The rotation period of the sun*
Determine the rotation period of the sun and the direction of its axis by observing the positional changes of sunspots. The materials required are: a small telescope, or binoculars (at least 6-power); a large box; a clip board; paper and pencil.

Warning. Do not look directly at the sun through the instrument. If binoculars are used, mount them firmly in the front end of a box (see diagram overleaf, *a*) but if a telescope is used, make a sunshade for it, as in *b*. One long side of the box should be left open for viewing. Prop the box up so that its rear end is perpendicular to the direction of the sun's rays. The clip board with paper attached is placed against the rear end, and the solar image is projected on to it. The eyepiece must be focused a little differently from the setting used for direct viewing (determine this by trial). Once the size of the solar image has been determined, all observations can be made of an image of this size, and paper can be prepared ahead of time by drawing a circle of proper size. Note that, with 6-power binoculars, a 5-centimetre solar image is obtained at 1 metre behind the eyepiece. Higher powers give proportionally larger images.

The image size is also proportional to distance from the eyepiece.

Make observations at the same time each day, preferably at noon. Always keep the paper oriented in the same way. In the circle, quickly mark in with pencil the locations of any sunspots. Then attempt to show their relative sizes and approximate shapes. It will be necessary to move the paper while doing this.

From day to day, the spots will appear to change position as the sun rotates. By measuring the differences between several daily sketches you can determine the rate of motion and, if the observations are continued for a month or more, you may see a spot group come back again the second time around. On the other hand, a large spot may disappear and new ones appear in the interim.

Observing the effect of the earth's motion

4.87 *A Foucault pendulum*
A G-clamp with a ball-bearing soldered to the inside of the jaw makes a good support for a Foucault pendulum which will show the rotation of the earth. It is best hung indoors with the ball-bearing resting on a stout razor blade or some other hard surface (see diagram).

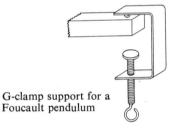

G-clamp support for a
Foucault pendulum

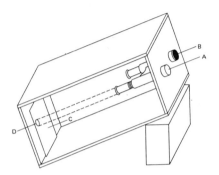

a using binoculars

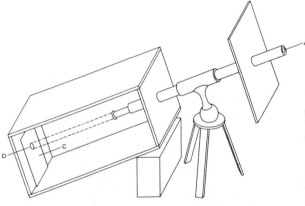

b using a telescope

Unspun nylon fishing line should be used for suspending the bob, which can be a solid rubber ball. The length of the pendulum is not important; anything from 3 m to 30 m will do.

When such a pendulum is set in motion, the plane of swing appears to alter after a few hours, as compared with a mark made on the ground at the time of release. It is, of course, the earth rotating underneath the 'bob' which gives this effect.

Care must be taken that the pointer, a short knitting needle driven into the ball, is continuous with the suspending thread. A reference line drawn on a piece of white card can be fastened to the floor with drawing pins. This must be positioned accurately under the pointer when the ball is at rest.

4.86   Observing the positional changes of sunspots:
     A to sun
     B shield over second opening
     C clipboard
     D sun's image

To set the pendulum in motion, attach a long cotton thread to a tin tack driven into the bob, and align it so that it lies along the direction of the reference line; then burn the thread near the tack.

It is not easy to get good quantitative results without many refinements, but it is not difficult to observe the effect.

4.88 *A miniature Foucault pendulum*

Mount a small Foucault pendulum from a stand set upon a turntable or office chair which can be rotated. Have the children observe the behaviour of the pendulum when the turntable is rotated slowly.

4.89 *The seasonal change of position of the sun*

A. From a fixed location with a good view, note accurately the point where the sun disappears behind landmarks as it sets. Repeat the observations at intervals of a week for four weeks at least, and determine the rate of change in degrees per day. (To measure degrees, a clenched fist at arm's length equals about 10°.)

B. Mark a line on the floor or the wall where the sun shines in your room and makes a shadow's edge. Note the exact month, day and hour. At the end of each week make another line at exactly the same hour. Repeat this throughout the year and you will have some interesting observations. The variation in position of the line from week to week and from month to month is caused by the movement of the earth around the sun.

C. In an open space, drive a 150 cm stick into the ground and let the children keep a record of the length of the shadow. Measure the shadow two or three times a day at different seasons of the year. The children should note the exact location of the stick's shadow at the same times each day, marking the length and position of the shadow. Get them to compare the position and lengths of the shadow at the beginning of the school year, during the winter, spring and at

end of the school year. (See also experiment 4.68).

D. Write a report explaining what the observed changes mean in terms of the motions of the earth.

4.90 *Photographing star trails*

A very interesting activity for pupils who have cameras is the photographing of star trails as the earth revolves. The materials required are a camera and film, a tripod or other solid support, and a watch or clock. Wait for a clear moonless night and find a place where there is an unobstructed view of the horizon. The place selected should be away from extraneous light such as automobile headlights, etc.

Face the camera as directly at your celestial pole as possible (Pole Star, if north of the equator), and secure it either with a tripod or with blocks of wood. Focus for infinity, open the diaphragm to full aperture, set the shutter for time exposure and start the exposure. Leave the set-up undisturbed for two hours or more, then close the shutter for a minute or two, taking care not to move the camera. Open the shutter again for a minute, and finally close it. This last short

Star trails around the north celestial pole

exposure identifies the end of the exposure. Record the times of beginning and ending.

On development, the film will show star trails as concentric arcs with centres at the celestial pole. The longer arcs can be measured to show how many degrees of rotation occurred and from this the period of full rotation can be calculated.

Similar exposures can be made with the camera pointing in different directions and at various elevations. Study the resulting trails and see how these all indicate a rotation of the entire sky as if it were a solid sphere with stars attached and the whole rotating about an axis through the celestial poles.

The moon's apparent path may be shown by taking exposures of 1 or 2 seconds each at intervals of 10 or 15 minutes during a couple of hours, or until the moon moves out of the field of the camera. Use extreme care to avoid displacing the camera.

By day, the sun's path can be recorded in the same way. *Warning*. Under no circumstances should you look at the sun through the viewfinder. Stop the lens down, to avoid excessive exposure. (See also experiment 4.75).

4.91 *Star trails in colour*
The stars are as colourful as land subjects, but this is not generally known because dark-adapted eyes have low sensitivity to colour. High-speed colour film and a camera with at least an f 3.5 lens will record the red star Betelgeuse in the constellation Orion, the yellow star Capella in the constellation Auriga, and the gold star Albireo in the constellation Cygnus. The constellation Cassiopeia contains two blue, one white, one golden, and one green star. A good camera that can make time exposures, a *rigid* tripod, and fast film are all you need. The simple star charts in this book will help you to identify the constellations. Your local public library may have books on amateur astronomy which contain similar charts. Dial indicators that show all the constellations overhead when the dial is set for the month, day,

and hour, are also obtainable in some countries.

The earth rotates 15° per hour, or 1° every 4 minutes. To us on the earth, it is easier to appreciate this movement by assuming that the stars move. Furthermore, the stars appear to rotate around your celestial pole. Each star near the pole traces a tight circle in its movement, and as the distance from the pole increases, the radius of curvature increases until the stars above the equator appear to travel in straight lines.

A star is a true point source of light and *no movement* of the camera can be tolerated unless you want pigtails for star images. All trouble can be avoided if you mount your camera on a rigid tripod, cover the lens with a cardboard, use a long cable release to open the shutter on time or bulb, wait 3 seconds or so for the camera to stop moving, and then remove the cardboard from in front of the lens. At the end of the exposure, again cover the lens with a cardboard before closing the shutter.

Note. Commercial processing laboratories will probably not recognize star images for what they are and, unless you instruct them otherwise, will return your negatives unprinted.

4.92 *Photographing constellations*
A. Photographs of constellations add an aesthetic purpose to photographing star trails. The results make beautiful prints and slides in both black-and-white and colour, and they prove to be a very effective teaching medium.

There are many techniques for photographing constellations, but a favourite is as follows: select a particular constellation, set up the camera, and expose for 30 minutes with high-speed black-and-white film (400 ASA) and a lens opening of f 11; then cover the lens for 2 minutes, open it to f 4, and throw it slightly out of focus; finally, uncover the lens for 3 more minutes. A diffusion screen over the lens for the final exposure works just as well as throwing the lens slightly out of focus. The resulting picture shows a constellation that appears to be plunging through space with a tail following each star.

B. Underexposed and discarded 35 mm film-slides can be perforated with a pinpoint in the form of various constellations. The slides can be projected on to a screen or used in a viewer, and pupils can try to identify the constellations. The slides can also be dropped into a slot made in a mailing tube and held up to the light (see sketch).

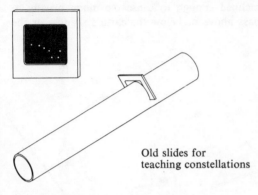

Old slides for
teaching constellations

4.93 *Photographing satellites*
Satellites are a joy to photograph. Use the same camera technique as for star trails (see above). Kodak Tri-X Pan film is an excellent choice. Use Kodak HC-110 developer, diluted 1 : 15 at 24° C for 4 minutes. The main problem is to know ahead of time where to aim your camera. There are several sources for this information: many newspapers publish daily the times, the degrees above the western or eastern horizon, and the direction of travel for all visible satellites. Also, local astronomical observatories and amateur astronomical clubs may be able to furnish the required data for you. Satellite photography is particularly rewarding when the satellite path crosses a well-known constellation, or if you are extremely lucky, perhaps two satellites will cross within your photograph. It is this unknown factor that continues to attract the amateur, as well as the professional, astronomical photographer.

4.94 *Finding the north-south line from the sun*
A. If you have a watch, set it to your local Mean (solar) Time, and proceed as follows:
North of the equator. Point the hour hand towards the sun. The north-south line is given by the bisector of the angle between the hour hand and 12 o'clock.
South of the equator. Point 12 o'clock towards the sun. The north-south line is found as above.

B. If you have no watch, you can use the shadow of a stick instead. Drive a stick vertically into the ground. As the sun crosses the sky in the course of the day, the shadow of the stick will turn. It will also grow shorter in the morning and longer again in the afternoon. When the shadow is shortest (close to noon) its far end will point north or south, depending on whether you are north or south of the equator.

Models and demonstrations for astronomy

4.95 *The phases of the moon and lunar eclipses*
The materials required are a focusing flashlight; a white ball and stand to hold it; an earth globe; and a darkened room.

A. With the flashlight fixed in position and shining full on the ball, have pupils view the ball from different directions to see crescent quarter phases, gibbous, and full 'moons'. Get them to write a report on these appearances as related to the changing phases in lighting of the real moon.
　　Rotate the globe to show how the times of rising and setting of the moon are closely related to the phase. For example, the first-quarter

The moon is high in the sky when observed from the point A on the earth's surface. At the same time an observer B would see the moon much lower in the sky

moon rises about noon, is highest in the sky at sunset, and sets about midnight.

By sighting across the position on the globe corresponding to your own geographic locality, you can simulate the relationship of the moon to the horizon for moonrise and moonset positions (see diagram on preceding page).

B. Eclipses may be demonstrated with the same apparatus. Placing the moon in the shadow cast by the earth globe simulates a lunar eclipse (which may be partial or total). Placing the moon between flashlight and globe will result in its shadow being cast on the earth. By this means, show that an eclipse of the sun is not visible over as great an area as an eclipse of the moon (which is seen from the entire half of the earth that is towards the moon). (See also the diagrams in experiments 4.84 and 4.85.)

The eclipse demonstration can also be adapted as an activity. All pupils could construct clay models of the earth and the moon, and illuminate them with flashlights.

4.96 *How solar eclipses appear*
The sun is represented by an opal electric bulb shining through a circular hole 5 cm in diameter in a piece of blackened cardboard. The corona is drawn in red crayon around this hole. The moon is a wooden ball, 2.5 cm diameter, mounted on a knitting needle. The observer views the eclipse through any of several large pin holes in a screen on the front of the apparatus (see diagram). The corona becomes visible only

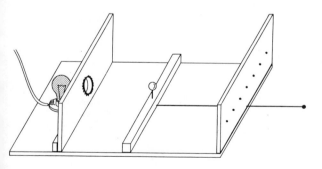

Simulating solar eclipses

at the position of total eclipse. The moon's position is adjusted by a stout wire bicycle spoke attached to the front of the apparatus. (See also experiment 4.84).

4.97 *Why an eclipse does not occur at every new and full moon*
The model is constructed as shown below using cardboard discs, and beads, marbles, ball bearings or modelling clay to represent the sun, earth and moon. The moon's orbit is inclined enough to cause the moon usually to pass above or below the earth's shadow or the

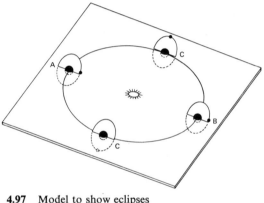

4.97 Model to show eclipses
 A eclipse of the sun
 B eclipse of the moon
 C no eclipse

region between the earth and the sun. Each small semicircular disc representing the moon's orbit is inclined in the same direction. Probably the inclination will have to be exaggerated to show the relation clearly. If desired, the cardboard sheet that represents the plane of the earth's orbit can be slotted and complete discs for the moon's orbit can be inserted, so as to show below as well as above the plane.

4.98 *The cause of seasons*
Use a hollow rubber ball such as a tennis ball to represent the earth. Push a 15 cm length of wire or a knitting needle through the ball to represent the earth's axis. Draw a circle about

40 cm in diameter on a piece of cardboard to represent the earth's orbit.

Hang an electric lamp about 15 cm above the centre of the cardboard to represent the sun. A lighted candle may also be used. Place the ball representing the earth successively at the four positions shown in the diagram with the axis slanted about 23.5°. Observe the amount of the ball that is always illuminated. Observe where the direct rays of the sun strike. In each of the four positions, observe which hemisphere receives the slanting rays of the sun.

Repeat the experiment with the needle

other through the centre. Where they cut the circle, label the intersections in counter-clockwise order: 20 March, 21 June, 23 September, 21 December. These are positions of the earth in relation to the sun on these dates. Draw a small circle for the earth at the 21 June position. The north pole will be off centre about $\frac{1}{8}$ radius of the circle, towards the sun. For any other date or orbital position (which can be located by using a protractor) the earth circle and pole will have the same orientation (see diagram). The Arctic circle, tropic of Cancer, and equator can be drawn in. Then a line through the centre

The differences in length of day and night
A 20 March   E Equator
B 21 June   F Tropic of Cancer
C 23 September   G Arctic circle
D 21 December   H sun

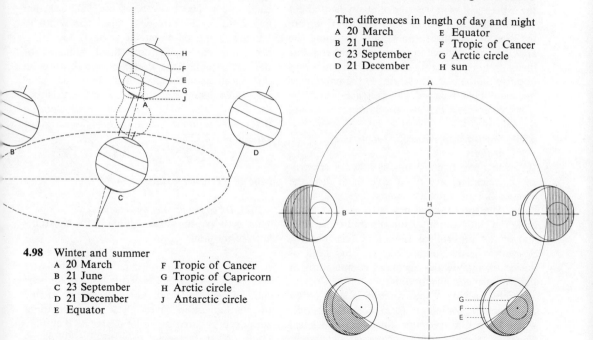

4.98 Winter and summer
A 20 March   F Tropic of Cancer
B 21 June   G Tropic of Capricorn
C 23 September   H Arctic circle
D 21 December   J Antarctic circle
E Equator

perpendicular to the table top in each of the four positions and observe what would happen if the axis of the earth were not inclined. (See also experiment 4.70A.)

4.99 *The causes of the differences in the length of day and night in some places*

Draw a large circle to represent the earth's orbit. Draw two lines perpendicular to each

of the earth circle and perpendicular to the earth–sun line will be the boundary between daylight and darkness.

From such a diagram, one can estimate the duration of sunlight at different latitudes for any date (e.g. on 1 August at the Arctic circle the sun would be estimated as up for about 18 hours according to the diagram; but up only 6 hours on 1 November).

4.100 *The effects of the angle of the sun's rays on the amount of heat and light received by the earth*

Bend a piece of cardboard and make a square tube 2 cm × 2 cm × 32 cm. Obtain a piece of very stiff cardboard and cut from this a strip 23 cm long and 2 cm wide. Paste this to one side of the tube with 15 cm extending. Rest the end of the stiff cardboard on the table and incline the tube at an angle of about 25°. Hold a flash-light or lighted candle at the upper end of the tube and mark off the area on the table that is covered by the light through the tube. Repeat the experiment with the tube at an angle of about 15°. Repeat again with the tube vertical. Compare the size of the three spots and determine the area of each. Show the analogy between this investigation and the way in which the sun's rays impinge on the earth's surface. Is the amount of heat and light received per unit area from the sun greater when the rays are slanting or direct?

4.101 *Making a spectroscope—materials analysis*

By using a sensitive instrument called a spectro-scope, scientists are often able to analyse the composition of materials located a great distance away. The spectroscope has been used to deter-mine the composition of the sun and other stars and of the atmosphere of many of the planets. Spacemen in the future will use this kind of device to analyse the chemical composition of their immediate surroundings.

Light entering a spectroscope is split up by a diffraction grating to form coloured bands, which we call a spectrum. Since each chemical element shows certain characteristic bright

lines in its spectrum the material can thus be easily identified.

The materials required are a shoe box, replica grating (see science supply catalogues), some masking tape, and a double-edged razor blade broken in two. Cut a hole of about 2 cm diameter in the middle of one end of the box. Use tape to fix a piece of replica grating over the hole from the inside. Cut a 2.5 cm × 0.5 cm slit, which should be parallel to the lines of the grating, in the middle of the other end. Cover the slit from the outside with a finer slit made from two halves of a razor blade, edges facing each other. The two halves are held together and fixed to the box with tape. The width of the slit should be about the same as the thick-ness of a razor blade and is finally adjusted for the best results (see diagram). Look through the spectroscope at various luminous gases such as neon and argon in lamps or signs. Notice the bright lines in the spectrum, which indicate that each element has its own pattern. (See experiment 2.222.)

Models for space science

4.102 *Discovering action-reaction*

This activity introduces children to Newton's laws of motion.

A. Have a child put on a pair of roller skates and throw a large ball over his head to another child. Does the child on roller skates move? In what direction? Why? Try the same thing with both children on roller skates as they play catch. What happens? (See also experiments 2.249, 2.250, 2.251.)

B. A rubber balloon can serve as a simple rocket engine. Obtain a balloon and let a child inflate it, hold it above his head with the end closed, and then release it. Let the children describe what happens.

This can be extended by letting the children inflate a balloon and tie up the end. Let them aim it at a target and attempt to hit the target.

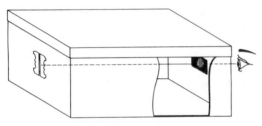

Shoe box spectroscope

Usually they are not successful because the balloon has no guidance system. Let them construct tail fins. These will affect the balance of the balloon and adjustment can be made by placing a rubber band around the front end of the balloon to which small weights such as paper clips can be attached. This gives the balloon balance plus a guidance system. Can they now hit the target?

4.103 *Building action-reaction engines*
A. *Balloon-powered boat.* Cut away one side of a milk carton and make a hole in the bottom near the edge. Next, insert a glass tube into the hole and attach the balloon to it with a rubber band (see diagram). Then inflate the balloon

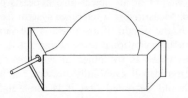

and place the boat in a tub of water. Does it move? In what direction? Does it make a difference in speed if the open end of the glass tube is under the surface of the water? If milk cartons of the type illustrated are not available, you can make a similar device out of a shoe box.

B. *Balloon-powered rocket.* Using cellophane tape, a child can attach a drinking straw to the side of a long balloon. Then he should pass a thin wire through the straw (see diagram) and attach one end of the wire to a school fence post or door handle. Next, get him to pull

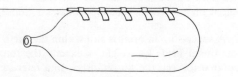

the wire taut and attach the other end to an object on the opposite side of the room or yard. After inflating the balloon he should release

it suddenly. How far will it travel? Use one variety of balloon and try different amounts of air. Quantify and graph the results of increasing the air. Repeat using various shaped balloons.

Perhaps the children can devise other types of machines in which there is a more efficient use of the action-reaction principle.

4.104 *Discovering thrust*
A. In the school playground or at home, children can begin to learn the meaning of thrust by feeling the thrust produced when water passes through a garden hose. As the amount of water passing through the hose is increased, the hose should begin to move. In what direction does it move? As the water pressure is increased, what happens to the motion of the hose? Attach a rotary lawn sprinkler to the hose. Gradually turn on more water. Notice the speed of the lawn sprinkler. Does it move faster or slower as the amount of water increases?

B. Thrust may be measured with a balance (see sketch). Put weights (10–50 g) on one pan. Firmly hold an inflated balloon over the other pan, then allow the air to escape against the pan. How many gm wt of thrust does the escaping air exert on the pan? (See also experiments 2.305 and 4.117.)

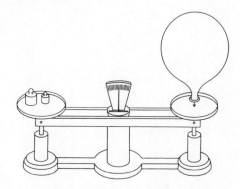

C. Large rockets may produce 300,000 to 1,000,000 kg wt of thrust. Develop with the pupils the idea that if a rocket weighs 5,000 kg, this means that the earth's gravity is pulling down on this rocket with a force of 5,000 kg wt. Before the rocket can rise, it must overcome that pull towards the centre of the earth. Therefore, the rocket's thrust must exceed 5,000 kg wt. Which of the following rockets would achieve the greatest height? Why?

Weight	Thrust
500,000 kg	500,500 kg wt
500,000 kg	750,000 kg wt
500,000 kg	1,000,000 kg wt

4.105 *Discovering weightlessness*
In order to study the motion of an object we need a reference system, e.g. something relative to which it is possible to describe the location of the object at any time. For many experiments we choose a reference system which is fixed to the earth, as for instance when we study a falling object. In such a reference system the earth is at rest. If we want to study the seasonal changes, however, we prefer a reference system where the sun is at rest and where the earth will be moving in an orbit. You see from this that the answer to the question whether an object is moving or not depends on what reference system we choose.

Not only the position but the weight of an

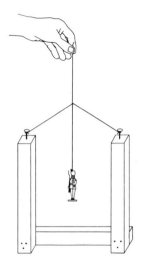

object depends on the reference system. The following experiment will demonstrate weightlessness.

A. Tie a string with a toy soldier or other object suspended from it loosely across the top of the three pieces of wood joined as shown in the diagram. Lift the entire apparatus and when it is hanging motionless release the string. While the soldier is falling, he can be seen to remain in the same position inside the frame. Since he is not supported by either the string or the frame, he is in a weightless condition with regard to his surroundings, e.g. the reference system being used.

B. The weight of an object also depends on its location. Measured in a reference system fixed to the earth, the weight of an object is the same as the earth's gravitational force acting on it. This force decreases as the object moves away from the earth and will eventually become negligible.

It should be noted that it is the weight of the object which is changing under the above circumstances. The mass (content of matter) of the object (measured in kg) does not change, unless we are dealing with relativistic physics, where objects experience speeds approaching that of light.

An astronaut whose mass on the surface of the earth is 90 kg still has the same mass of 90 kg on the surface of the moon but his weight, which is 90 kg wt on the earth's surface, would only be about 15 kg wt on the moon's surface. Using SI units, the mass is m kg but the weight is mg Newton. Since g at the moon is approximately one-sixth of g at the earth, the weight of a man on the moon will be one-sixth of his weight on the earth. (See Appendixes 1 and 2).

C. A space-ship in orbit is still within the earth's gravitional field. Its weight is exactly the force required to keep the ship in orbit. In a reference system attached to the ship, however, everything inside the ship is weightless and with a slight push against one wall of the cabin a man can propel himself towards the opposite wall.

Further away from the earth, the gravitational force becomes negligible and the space-ship will move in a straight line unless acted upon by forces from its own engine or from other objects like the moon (Newton's first law). Outside the space-ship a man could, if he were completely free to move, push himself off in any direction never to return. To avoid such a possibility, safety lines are attached to the space suits of astronauts who work in space.

4.106 *A satellite launcher*
Materials required are a bucket, a football, a coathanger (or other suitable wire), sinker or weight, a piece of string and a test-tube or a cap of some sort.

Place the ball securely in the bucket. Bend the wire so that about 30 cm of it is straight and the rest is curved into a circular base as shown in the sketch. Using masking tape, secure the circular portion on the ball, allowing the straight, 30 cm portion to stand upright in the centre of the top of the ball. Attach the sinker or weight to the string. Fasten the other end of the string to the test-tube or cap with tape. Invert the cap on top of the upright wire (see diagram).

Explain that the ball represents the earth, and the sinker represents the satellite. All that it

takes to set the sinker into motion in any direction is the tap of a finger. Let the children find out what happens when the satellite is launched in the following different ways:

1. With a slight tap, push the sinker up and away from the surface of the ball, as shown in the figure. What happens? (The sinker moves up and then falls back to the starting point. This is how an object travels when it is projected at low speed straight up from the earth.)

2. With a slight tap, push the sinker off the surface of the ball at an angle. Show by a diagram what happens. (The sinker moves away from the ball and then falls back at some distance from the starting point. The distance spanned depends upon the angle of launching and upon the forcefulness of the tap.)

3. With a stronger tap, push the sinker off the surface of the ball at an angle. Make a diagram of the orbit. (The sinker moves away from the ball, circles it, and lands. Evidently, a complete orbit passes through the starting point of the orbit.)

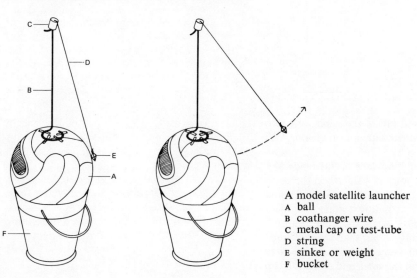

A model satellite launcher
A ball
B coathanger wire
C metal cap or test-tube
D string
E sinker or weight
F bucket

Weather

Making weather instruments and a weather station

Weather is a topic that is close to the life of every child. Even at the lowest levels of primary instruction, observations of the weather may be made from day to day. At the intermediate levels a simple weather station may be constructed. At the level of general science and later, a more detailed study of the causes of weather phenomena may be made. At all stages of the work it is an advantage to represent readings and observations in graphical form whenever this is possible.

4.107 *Making a wind vane*

A wind vane is used to find the direction of the wind. Obtain a piece of wood about 25 cm in length and 1 cm square in cross-section. With a saw cut a slot, 6 cm deep, in the centre of each end of the stick.

Next select a thin piece of wood about 10 cm wide which will fit tightly in the slots. From this cut two sections, one the head of an arrow and the other the tail, as shown in the figure.

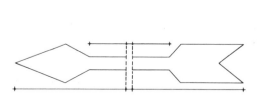

Push the head and tail of the wind vane into the slots and fasten them either with glue or with small nails.

Next balance the wind vane on the blade of a knife and mark the point of balance on the stick. Take the glass part of a medicine dropper and close the small end by rotating it in a gas or alcohol flame. At the place where the vane balanced, drill a hole just slightly larger than the medicine-dropper tube about three-quarters of the way through the stick. Put the small end of the tube up

in the hole and fasten it securely with glue or putty.

Make a supporting rod for your wind vane by selecting a piece of soft wood about a metre in length and driving a small nail in the top. With a file, sharpen the end of the nail to a point. Place the medicine dropper over the nail and mount your wind vane on top of a building or on a pole so that it is exposed to the wind no matter from which direction it blows.

Fix stout wire arms to the pole and bend the symbols N, E, S, W, at the ends, or solder large letters cut from sheet metal to each free end.

4.108 *Making a wind speed indicator*

Select two pieces of light wood about 50 cm long and 1 cm square in cross-section. Cut a notch 1 cm wide and about 0.5 cm deep at the exact centre of each piece, and assemble as shown in the diagram.

Obtain the glass tube from a medicine dropper and close the small end by rotating in a gas or alcohol flame. As in the preceding experiment, drill a hole about three-quarters of the way through

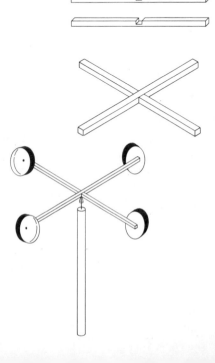

the wood at the exact centre of the cross arms and set the medicine-dropper tube securely in the hole with cement or putty. Obtain four cigarette tins or small plastic dishes and fasten them to the ends of the cross arms with small nails or screws. Make sure the cups are all facing in the same direction (see figure). Prepare a mounting stick for the wind indicator in exactly the same way as you did for the wind vane. Drive a nail in the end of the stick and sharpen it to a point with a file.

Your wind speed indicator will spin in the wind. You can get a rough idea of the speed of the wind in kilometres per hour by counting the number of turns made in 30 seconds and dividing by 3.

Another way of determining the wind velocity is to get someone to drive you in a car on a calm day. Hold your speed indicator out of the front window and ask the driver to go steadily at 5 kilometres per hour. Count the number of turns in 30 seconds for this speed. Repeat with the driver going at 10, 15, 20, 25, 30, 40, etc., kilometres per hour.

Mount your wind speed indicator in a place that is exposed to the wind from any direction.

4.109 *Making a deflexion anemometer*
To make a deflexion anemometer, first obtain a piece of wood measuring approximately 25 × 2 × 1 cm. Make a saw cut at one end into which a protractor can be inserted. Use a little glue to attach the protractor securely. Before the glue sets, drill a hole 0.5 cm in diameter through the stick at the centre of the protractor. Bend a short length of coat-hanger wire as shown in the figure and hang it in the hole you have just drilled. Cut a piece of cardboard so that it measures approximately 10 × 8 cm and fasten it to the wire with sticky tape or staples. Note that the cardboard vane is slotted so that it will swing around the protractor as the wind blows against it.

Now balance this portion of the anemometer on the straight edge of a ruler by fastening wood screws to the lighter end as needed. At the balance point, drill a hole large enough to accept a glass tube bearing. The bearing can be made by holding the end of a length of glass tubing in a flame long

enough for the opening to become sealed. When cool, cut off 3 cm of the sealed end; the result will resemble half of a medicine capsule. Insert the bearing in the drilled hole and glue firmly in place.

Next, drive a nail into the end of a piece of broomstick or similar wood, cut the nailhead off and file a sharp point on the cut end. Set the anemometer on this nail so the glass bearing is resting on the point. It should pivot freely and face into the wind (see diagram).

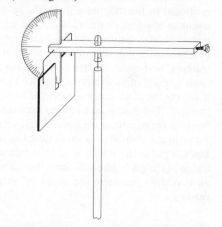

To calibrate this device, ask someone to drive you along a side road on a calm day. Hold the anemometer out of the window and as the car moves along at 5 kilometres an hour, make a mark on the protractor where the vane hangs. Repeat for speeds of 10, 15, 20, and 25, etc., kilometres per hour. While not a precision instrument, this anemometer will give a surprisingly good performance.

4.110 *Making a pressure tube anemometer*
The principle upon which this type of anemometer operates can be demonstrated by supporting a funnel attached to a U-tube near a three-speed fan.

An effective instrument may be constructed by pupils who are interested in manual-type projects. A U-tube, with funnel attached, is fitted to a flat board as shown overleaf. A vane made of thin wood or metal is screwed to the board to keep the mouth of the funnel facing directly into the wind.

A ruler also fixed to the board measures the height to which the water rises in the arm of the U-tube. Two screw eyes are fitted into the back of the board and these pass over a dowel (a piece of wood with a circular cross-section) so that the lower eye rests on a nail driven through the dowel. Lubricate the points of contact between the screw-eyes and the dowel with vaseline or other lubricant (see side view, below).

The speed of the wind is approximately proportional to the difference in height between the columns of water in the U-tube. The amount of water initially required in the U-tube is found by experimentation during the calibration of the instrument, and once this has been determined the same amount of water must always be maintained if any degree of accuracy is expected from the instrument. The instrument can be exposed to sub-freezing temperatures if alcohol is substituted for water in the U-tube. If a small quantity of lamp-black is put in the water, a 'dirty' ring round the inside of the glass tube will show the highest wind gust velocity between the times of instrument reading.

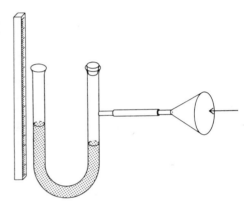

4.110   A pressure tube anemometer

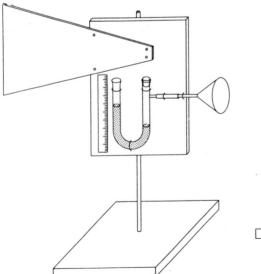

Side view, instrument separated to show the two screw-eyes and supporting nail

4.111 *Making a rain gauge*
A. *A simple rain gauge.* It is easy to make a simple rain gauge using a funnel and bottle, with a measuring cylinder to measure the volume of water collected in the bottle (see figure). Ideally, the funnel should have either a very sharp vertical edge or a horizontal lip to prevent raindrops bouncing out again. The whole apparatus should be buried so that the funnel is a few centimetres above ground level.

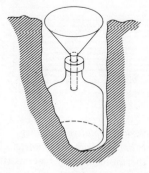

B. *Another rain gauge.* Procure a large tin can about 10 cm in diameter and 14 cm in height. Next secure a straight-sided glass vessel about 3 cm in diameter and at least 25 cm high, that will stand inside the can (see figure). Place the can on a level table and pour water into it until the water

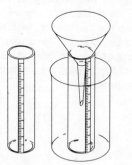

is exactly 1 cm deep on a ruler. Paste a strip of paper about 1 cm wide along the length of the tall straight-sided jar. Next pour the water from the can into the tall jar and make a mark on the paper strip at the level where the 1 cm of water from the

can comes. Measure the distance from the inside bottom of the tall jar to this mark and mark off equal spaces along the strip of paper to the top. Divide the distances between the marks into 10 equal parts to measure millimetres. The jar will measure small amounts of rainfall.

To assemble the rain gauge, select a funnel whose diameter is approximately equal to that of the tin can and place it in the tall jar. Stand both in the tin can. Set the rain gauge in an open spot where it will not be upset easily.

If the rainfall is light it can be measured by the small jar alone. If it is heavy, excess water will overflow into the larger can and may then be measured by pouring it into the bottle. If the rainfall is to be measured in inches, pour 2.5 cm of water in the large can and then pour this into the tall jar. Mark the depth to which the 2.5 cm of water reaches and then divide the scale accordingly.

A better way to determine the rainfall in centimetres or inches is to graduate the smaller measuring bottle in terms of its radius and the radius of the collecting funnel by use of the formula:

$$\frac{\text{Height in bottle for each}}{\text{cm of rainfall}} = \left(\frac{\text{Radius of funnel}}{\text{Radius of bottle}}\right)^2$$

4.112 *Making a hair hygrometer*
This device will enable you to read the relative humidity directly without the use of tables.

Procure a few human hairs about 30 cm long. Free them from grease by dipping them in dilute caustic soda solution. Attach one hair to the upper end of a stand and stretch it with a 50 g weight. The hair should pass two or three times round a spool fixed to an axle which is free to rotate in bearings made from a piece of tin and fastened two-thirds of the way down the stand. For greatest sensitivity the diameter of the spool should be small. Fix a light pointer of balsa wood to the axle, and arrange a postcard to act as a scale (see figure overleaf).

Changes in atmospheric humidity will affect the length of the hair and the position of the pointer.

To mark off the scale it is best to compare your hygrometer with a standard one. If one of these is

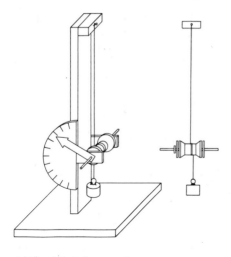

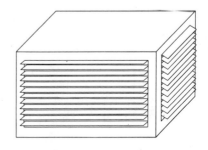

4.112   A hair hygrometer

sides form a roof and another a floor for your house. The open side and the two ends should be fitted with louvres such as are found in a window blind, for best results (see figure). This will provide a free access of air but will protect the instruments from wind and precipitation.

not available, place the instrument above some warm water in a pail and cover with a wet towel. When the pointer has moved as far as it will, mark this point 100 on your scale for the air in the pail will be 100 per cent saturated. Other points can be marked by taking readings on your wet and dry bulb hygrometer (see experiment 4.114). Find the relative humidity from table, Appendix 6, and mark the position of the pointer on your scale accordingly. When you have established about three points on your scale you can then divide the rest into equal divisions and mark them off at 5-interval markings from 5 to 100.

4.113 *Making a housing box for weather instruments*

Some of your weather instruments must be exposed to the weather. Among these are the wind vane, the wind speed indicator and the rain gauge. It is wise to protect the metal parts of these instruments with either grease or paint. Aluminium paint also works very well for this purpose.

Other instruments such as the barometer, the thermometer, and the hygrometer, need to be shielded from rain and wind. These may be placed in a wooden box which has no top. Place the instruments in the box so that one of the closed

4.114 *Measuring relative humidity*

A. *The wet and dry bulb thermometer.* The wet and dry bulb thermometer may be constructed easily. Two thermometers that have the same reading under similar conditions are securely mounted on a piece of wood. Sew together a strip of muslin to make a snug-fitting 'sock' over the bulb of one thermometer. (The sock or wick may be purchased ready for immediate attachment.) Mount a small narrow-mouthed bottle to the board so that the top of the bottle is at the same level or slightly lower than the top of the bulb (see figure). Keep the bottle filled with water. Before taking a reading, fan the air across the wet bulb for a minute or two. In order to determine the relative humidity, consult the psychrometric tables given in Appendix 6.

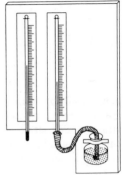

B. *The sling psychrometer.* If a sling psychrometer is not available, the instrument shown in A preceding may be converted into one by boring a hole in the top of the board, adding a strong rope and removing the reservoir of water. When it is spun around in the air, maximum evaporation occurs and more accurate readings are possible. The thermometers must be securely mounted before the instruments are swung. Pupils should also be instructed how to swing the device since striking their bodies or a desk are common accidents resulting in broken thermometers. Using the psychrometric tables (Appendix 6), calculate relative humidity of air both inside and outside the school and account for differences, if any.

C. *The dewpoint hygrometer.* The dewpoint hygrometer consists of a brightly polished metal cup and an accurate thermometer suspended in the water partially filling the cup. Hold the thermometer in the cup by inserting it in a pencil clip attached to the inside of the cup. Put a cube of ice in the water and stir. Continue stirring until the first evidence of dew appears on the outside of the cup. Immediately read the temperature of the chilled water (the dewpoint temperature) and the atmospheric temperature. Use the psychrometric tables in Appendix 6 to determine the relative humidity. (See also experiments 4.112 and 4.134.)

Winds and weather

4.115 *Air expands when heated*
To show that air expands when heated, fit a bottle with a one-hole stopper or cork which has a 30 cm length of glass tubing or a soda straw through it. Place the end of the tube in a small bottle of water. Heat the flask and observe what happens. Heat the flask until a considerable amount of air has been removed and then cool the flask by pouring cold water over it or by rubbing it with a piece of ice. What do you observe? How do you account for this?

Here is another way to show that air expands when heated. Fit a toy balloon over the neck of a small bottle and place the bottle in a pan of warm

water. What do you observe? How do you account for this?

Air exerts a pressure because it has weight; this pressure can be measured by a barometer. The three most important factors which affect air pressure are altitude, temperature and moisture. (See also experiment 2.110.)

4.116 *Air has mass*
The fact that the atmosphere has mass can be demonstrated quickly and inexpensively with a large balloon. Inflate it fully, place it on a platform balance and find its mass. Carefully remove the balloon without disturbing the balance or its weights in any way. Deflate the balloon and replace on the balance pan. Pupils will observe that the balance is no longer 'balanced' and that the side with the balloon has become lighter. (The effect of buoyancy has been disregarded.) (See also experiment 2.304.)

4.117 *Air exerts pressure*
Atmospheric pressure may be demonstrated in many ways. Atmospheric pressure causes liquids to rise in a sipping straw. A flask or bottle is prepared with a 'straw' of glass tubing and a short glass right-angle elbow held in a rubber stopper, as shown in the diagram. When the end of the bent tube is closed with the finger it is difficult to sip liquid up through the straw, but it is easy when the finger is removed. To demonstrate that pressure on the surface of the water is the factor that causes the liquid to rise in the tube, the pressure can be raised by blowing through the right-angle tube. For a variation of this demonstration,

a flask is completely filled with water and closed with a rubber stopper containing a simple length of glass tubing. A pupil can be challenged to drink the water through the 'straw'. If air is completely excluded from the bottle, he will be unsuccessful. (See also experiments 2.305 and 4.104B.)

4.118 *Cold air is heavier than warm air*
A. Obtain two paper bags that are the same size. Open the bags and attach a 20 cm thread to the bottom of each one with a piece of cellulose tape or by making a hole in the bottom of the bag, inserting the thread and then tying a knot in the end. Make a loop on the other end of each thread that will go over the ends of a balance rod as shown in the figure. Place a bag near each end of the rod. Move the bags in or out until they are in exact balance. With a candle, heat the air well below one of the bags. What do you observe? Let the balance stand for several minutes. What happens? Now heat the air under the other bag. Observe what happens. How do you account for this?

B. Another way to study the difference in mass between equal volumes of warm and cool air is to use flasks on the balance rather than paper bags. Attach the flasks with loops of string. Move them until they are in perfect balance and then heat one flask gently. Observe the effect. Allow to cool to room temperature. Observe and then heat the other flask. Flasks made from old light bulbs work very well for this experiment.

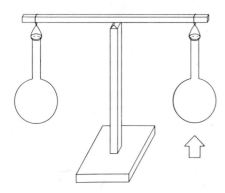

4.119 *Making a convection box*
A box to show why winds blow may be made easily. Obtain an open wood or pasteboard box and cut a pane of glass so that it just covers the opening of the box and makes a window. A wood chalk box which has grooves for a cover works very well. Cut the glass so that it will slide in the grooves (see figure). Next bore two holes in one of the long sides of the box, one near each end. The

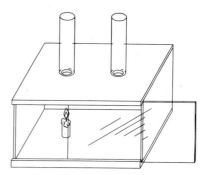

holes should be from 2.5 to 3 cm in diameter. The box must lie with this side up. Get two lamp chimneys to place over the holes. If lamp chimneys are not available you can use pieces of mailing tube about 15 cm in length. Place a short piece of candle on the floor of the box just under one of the chimneys. Light the candle. This represents a land area that has been heated by the sun. Close the window and with a piece of smouldering paper, trace the air current in each chimney. Observe the movement of air inside the box. Move the candle so that it is under the other chimney and repeat the experiment. What do you observe? How do you account for this? This is called a convection current. (See also experiment 2.128.)

4.120 *Tracing convection currents*
A. Shield a burning candle to protect it from stray air currents. Trace the air currents about it with smouldering paper.

B. Open a door a little way between a warm and a cool room. With a piece of smouldering paper explore the air currents about the opening at various levels above the floor.

C. If you can, explore the air currents in a room that is heated with a radiator or a stove.

D. Explore the air currents in a room that is ventilated with windows open at the top end and at the bottom.

E. Lower a lighted candle into a milk bottle by means of a wire. Observe what happens. Venti-

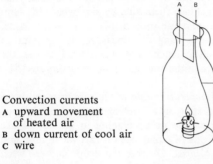

Convection currents
A upward movement
   of heated air
B down current of cool air
C wire

late the bottle with fresh air. Again place the lighted candle in the bottle but this time separate the warm and cold air currents by a piece of cardboard cut in the shape of the letter 'T' as shown in the diagram. Explore the air currents on each side of the cardboard with smouldering paper.

F. Cut out the top of a can so as to make a metal disc. Punch a depression in the exact centre. Cut along radial lines almost to the centre and give

Using convection currents
to turn a wheel

each of the blades thus formed a twist in the same direction (see figure). Mount the wheel on a pointed wire and hold it over a candle or other source of heat. A carefully made wheel of this kind will also turn over a radiator or a lighted electric lamp. (See also experiment 2.127.)

G. A more sensitive wheel can be made from the metal foil top of a milk bottle. Place the top on a piece of blotting paper with the flat side down. Press the point of a ball pen into the middle to make a dent. Cut 'petals' in the turned-up edge to form the vanes of a turbine. Pivot it on a pointed wire or a needle stuck into a cork.

How moisture gets into the air

4.121 *Atmospheric moisture*
You cannot see atmospheric moisture. This can be demonstrated in the following way.
    Put some water in a kettle or similar vessel and place the kettle over a fire. If a kettle is not available, fit a flask with a one-hole stopper and place a right-angle bend of glass tubing in it. Place some water in the flask and put it over a flame. When the water is boiling and steam issues from the spout, observe the cloud that is formed. This is not

steam, but condensed water. Observe the space next to the spout where the steam comes out. Can you see the steam? Now hold a candle or a burner in the cloud of condensed steam (see figure). What do you observe? Where does the moisture go?

4.122 *Weighing water 'lost' by evaporation*
Wet a bath towel with water and wring it out.

Weighing water 'lost' by evaporation

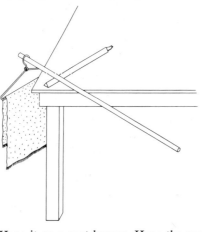

Hang it on a coat hanger. Hang the coat hanger on one end of a long stick balanced on a triangular file which is lying on a corner of a table. Look at the towel an hour later. Why has it lost weight? Where has the water gone? Hang weights on the hanger until the balance is restored. How much water has evaporated?

4.123 *Moisture evaporates from soil*
Fill a flower pot with moist soil and place it on a pair of scales. Either balance the pot of soil with weights or observe its mass. Observe its mass again after 24 hours.

4.124 *Moisture comes from house plants*
Place a cellophane bag over a leaf of some house plant or garden plant and close the end about the stem with a rubber band. Make an observation after one hour. What do you see? Where did the moisture come from?

4.125 *Moisture comes from other plants*
Plant some bean or pea seedlings in a flower pot and let them grow until they are 10 or 15 cm in height. Cover the top of the pot with cellophane or sheet rubber, pinning it closely around the stems of the plants so that no soil is left uncovered. Invert a clean, dry glass jar over the plants and observe after an hour. What do you see? Where did the moisture come from?

4.126 *Moisture from breathing*
Moisture coming from breathing may be shown by blowing on a cool mirror or into a cool glass or bottle.

4.127 *Surface area affects the amount of evaporation*
Obtain a large flat dish such as a baking tin and half fill it with water. Put the same amount of water in a tall can with a smaller diameter than the dish. Place them side by side where the temperature and air movements will be the same. On the following day measure the amount of water left in each container. What causes the difference in the amount of water evaporated?

4.128 *Temperature affects the rate of evaporation*
Warm a spot on a level blackboard or slate by using a candle or by placing in the sun. Place water drops of equal size on this warm area and on a cool area. Observe the drops and see what happens.

4.129 *Moving air affects the rate of evaporation*
With a moist sponge or cloth, wet two areas of equal size some distance apart on a cool blackboard surface. Fan one area with a piece of cardboard and leave the other to evaporate without fanning. What causes the difference in rate of evaporation?

4.130 *Moisture in the air affects the rate of evaporation*
Fasten some cloth over a wooden hoop or frame that is about 30 cm square and about 3 cm thick. Wet the cloth. Next make two wet spots on a cool blackboard surface with a sponge or cloth. Cover one with the frame carrying the wet cloth and leave the other one open. After a few moments observe both spots. Which has evaporated the faster? How does moist air (under the frame) affect the evaporation of the water from the spot under the frame?

How moisture comes out of the air

4.131 *Moisture condenses on cool surfaces*
Place some ice in a shiny tin can. After a little while observe the outside of the can. What do you see? Where did it come from?

4.132 *Investigating the water cycle*
Heat some water until it is near the boiling point. Place it in a drinking glass and rotate the glass so as to moisten the sides right to the top. Put some very cold water in a round flask and place the flask on the glass at an angle as shown in the figure.

Water will evaporate from the hot water, condense on the cool surface of the flask and fall back in droplets into the glass. Here you have evaporation, condensation and precipitation. You have seen the water cycle as it is in nature.

4.133 *Reproducing the rain cycle*
Place a box of plant seedlings on a table and over it fix a metal tray so that the distance between the box and the tray is about 35 to 40 cm. Put some

pieces of cracked ice in the tray. Place a tea kettle or flask containing water over a source of heat so that steam will issue between the seedlings and the tray (see the figure). You are now ready to study the rain cycle in miniature. The tea kettle or flask serves as the earth source of water. This evaporates and rises up to the cool tray which represents the upper layers of air above the earth and which have been cooled by expansion. The moisture condenses on the tray and drips back on to the seedlings as rain.

4.134 *Dew-point temperature*
You can measure the dew-point temperature with a shiny can containing some water, a thermometer and some ice. The dew-point temperature is an important weather observation. It is the temperature at which the moisture in the air begins to condense. The dew-point temperature changes from day to day.

Be sure that the outside of the can is clean and free from fingerprints. Stand the can on a page of printing so that the printing is clearly reflected from the can. Now add ice, a little at a time, to the water and carefully stir with the thermometer. Keep a close watch of the temperature and read the thermometer at the temperature where dew begins to form on the outside of the can, that is, when the print is no longer clearly visible. This will be near the dew-point temperature. (See also experiment 4.114).

4.135 *Making cloud in a bottle*
You can make a cloud form in a bottle. Obtain a large glass bottle and fit it with a rubber stopper carrying a 10 cm length of glass tubing. Place about 3 cm of warm water in the bottle and shake a little chalk dust into the air inside. Connect the glass tube to a bicycle pump with a piece of rubber tubing (see overleaf). Hold the stopper in the bottle and have a pupil pump air in. When the air has been compressed inside the bottle, let the stopper blow out and observe what happens. If you do not get a good cloud, introduce a little smoke from a smouldering match or piece of paper.

When the air expands it cools, thus reducing

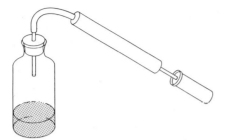

4.135   Making cloud in a bottle

the temperature in the bottle below the dew point. The moisture then condenses and forms a cloud. Similarly, when warm air rises above the earth the air pressure is reduced. The air expands and cools, and clouds form when the cooling goes below the dew point.

4.136 *Studying snow-flakes*

If you live in a region where snow falls, collect some snow-flakes on a piece of dark wool cloth and look at them with a magnifying glass (see the figure). You will find them of many, many shapes, but always six-sided. Snow-flakes are among the most beautiful sights in nature.

Weather projects

4.137 *Keeping a weather record*

The date, hour, temperature, sky and wind can all be recorded in a table and the readings should be

taken at the same time each day. Entries can be made in a note book under the headings:

Date	Time	Temperature	Sky	Wind	Rain

Useful information can be obtained by making graphs of temperature/time, rainfall/time, change in appearance of the sky over a period of time and also changes in wind intensity.

    Abbreviated scales can be used unless the records are for official purposes when recognized international symbols should be written. If no thermometer is available, a suitable temperature scale is: hot, warm, moderate, cool, cold, very cold. The velocity of the wind can also be recorded, as follows :

    Light—moves smoke, but not wind vanes.

    Moderate—raises dust and just moves twigs.

    Strong—large branches move.

    High—blows dust, papers; moves whole trees.

    Gale—breaks off twigs from trees.

The direction of the wind can be indicated by an arrow in the wind column of your records, but it is also possible to construct a paper star as shown in the diagram and to draw a line each day along the arm which most nearly coincides with the direction of the wind. The other symbols can be used to indicate general conditions.

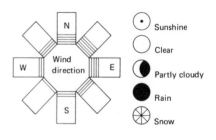

4.138 *Miniature weather fronts*

The following demonstration can be used effectively to illustrate the concept of what happens when two weather fronts come together. The directions given below should be followed exactly and the aquarium prepared carefully:

(a) Prepare an aquarium (any size) by gluing plastic rib binders, such as the type from acetate folders, along the bottom and sides of the aquarium. The folders are the type used by students to hold theme papers together and can be purchased from a stationery store. Plastic model cement or contact cement may be used to glue the plastic rib to the inside of the aquarium sides and floor. This forms watertight and waterproof guides into which the glass partition can be inserted. Metal guides kept in place by putty can be used instead of plastic binders. Make a partition from a piece of glass cut large enough so that it will remain in the plastic guides, but still clear the top of the aquarium (see figure *a* below).

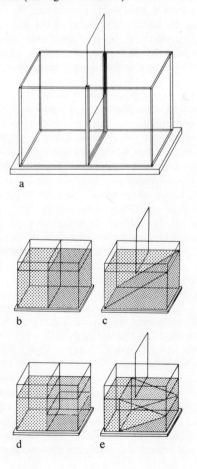

a

b   c

d   e

(b) Put warm water in one compartment and cool water in the other. Add red food colouring to the warm water and blue colouring along with some salt to the cool water.

(c) Remove the partition: Blue (representing a cold air mass) will sink to the bottom and red (representing a warm air mass) will be on top. This will be striated without much mixing.

(d) To demonstrate an occluded weather front, reinsert the partition. Stir one side gently (intermediate mass).

(e) Remove the partition. The intermediate mass (occluded weather front) will force its way between the warm and cold layers to form three distinct layers.

Much independent investigation can result when youngsters start asking and finding out answers to questions such as: What will happen if no salt is used? What will happen if the two coloured waters are the same temperature? What is the temperature difference between the two (or three) layers of water?

4.139 *Measuring the upper winds*
Materials required are: a balloon filled with 'lighter-than-air' gas, two protractors, metric ruler, a piece of wood 40 × 2 × 2 cm, a weight, some tacks, a watch with a second hand, and a spool of white thread.

The procedure is as follows. Tack a protractor to the side of the wooden stick as shown overleaf, keeping the straight edge of the protractor parallel to the top of the stick. Hang the weight on a piece of thread from the centre of the protractor to serve as a plumb bob. (A soda straw taped to the top of the stick will improve the sighting.) You now have a simple hand transit.

When the plumb bob indicates 90° on the protractor, the transit is horizontal. When the plumb bob indicates 80°, the transit is inclined 10°. An indicated angle on the protractor must be subtracted from 90° to find the inclination of the transit.

Practice in the use of the transit in the classroom is helpful. Ask a child to stand a measured distance (3 to 5 metres) from the wall of the classroom, and from that point find the angle that his

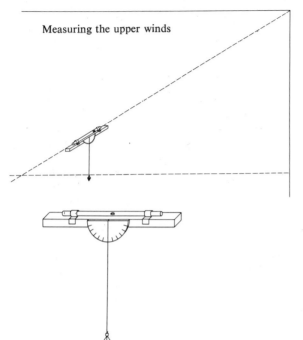

Measuring the upper winds

line-of-sight to the top of the wall makes with the horizontal. He can do this by finding how many degrees above the horizontal the soda straw had to be elevated in order to sight the top of the wall. On a graph paper, the child should measure off horizontally the number of units equivalent to his distance from the wall. At the end of this horizotal distance, he should copy the angle of elevation indicated by his transit. His scale drawing of his distance from the wall, and his angle of elevation, will indicate the height of the ceiling above his eye level. For example, suppose a child is standing 7 metres from the wall and finds the top of the wall to be 30° above the horizontal (a protractor reading of 60°): the ceiling would be nearly 3.5 metres above his eye level. To find the height of the room the child's eye-height is added to the 3.5 metres.

Now tie a long thread to the gas-filled balloon so it can be pulled down if released in the high-ceiling room. Pull the balloon to the floor, release it and measure the time it takes for the balloon to

strike the ceiling. Do this several times to get the average rate of ascent from floor to ceiling. Divide the ceiling height by the time of rise of the balloon to find its rate of ascent.

Now the children can take the balloon out of doors to measure the upper winds. One child should be assigned to each of the following tasks: (a) keeping the balloon in sight through the soda straw; (b) reading the angle of the plumb bob every 30 seconds; (c) keeping time, calling off each 30 seconds to the angle-reader; (d) writing down the elapsed time and the angle of sight at the end of each time interval.

When the data for a few minutes' sighting has been recorded, the position of the balloon at the end of each 30-second interval can be plotted. When the balloon's position at the end of each time interval has been plotted, its horizontal movement can be measured using the same scale for both vertical and horizontal distance.

4.140 Constructing a weather picture
Immerse a piece of white blotting paper in a solution containing two parts cobalt chloride to one part common salt. While wet the paper will remain pink, but when dried in the sun or near a bunsen burner it turns blue. This is the basis of the weather pictures.

A home-made picture can be made in the following way. Obtain a picture in which there is some sky or water and make an inset of the prepared blotting paper to replace the sea or the sky. The picture should then be mounted on a card and hung near a window where it will quickly respond to changes in the hygrometric state of the atmosphere.

4.141 Measuring the amount of dust in the air
To measure the amount of dust that falls in your neighbourhood, you will need at least three wide-mouthed glass jars, 5-litre size. You will also need about 10 litres of distilled water. (Ordinary tap water may contain tiny particles that would affect your measurements of dustfall.) You will also need a 2- or 3-litre pan or other container that can be heated without breaking. Last but not least, you will need to use a balance that weighs things

to the nearest centigramme or milligramme.

Make sure that the jars are clean, then rinse them out with some of the distilled water. Next pour 1.5 litres of distilled water into each jar. Mark the water level with finger-nail polish, a file mark, or anything else that rain won't wash away. Cover the top with a wire screen to keep

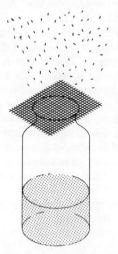

insects out. Put each of the jars in a different location outdoors. They should be about 1.5 metres above the ground, and not under trees or eaves of buildings.

Leave the jars in their places for 30 days. Visit them every few days and add distilled water to the original water level. (If the jar dries out, the wind may blow away the dust.) Rain may fall into the jar, but this causes no trouble unless the jar overflows: if this happens, the experiment will have to be repeated.

After 30 days, bring the jars indoors. To find out how much dust is in each jar, first weigh the 2- or 3-litre pan on the balance and write down the result, then pour the water from the jar into the pan. Use more distilled water to rinse out the jar, making sure all the dust particles have been removed. Then, heat the water until it all evaporates. Don't overheat the pan or the dust will be burnt.

Let the pan cool and then weigh it on the balance. This gives the weight of the dust and the pan. The weight of the dust alone can be found by subtracting the weight of the pan, which you have already noted. If the balance used only weighs in centigrammes, multiply the weight of the dust by 10 to change it to milligrammes.

Your figures tell you only how many milligrammes of dust fell through the mouth of the jar in a month. If you want to know how many metric tons per square kilometre this is equal to, first find the area of the mouth of the jar in square centimetres. Simple division will give the fall in milligrammes per square centimetre. If this answer is multiplied by 10 you will have the number of metric tons per square kilometre. (If the fall is required in tons per square mile the answer can be found quickly by multiplying the number of milligrammes per square centimetre by 25.5.)

Do you get the same sort of figure for each of the jars you set out? (If the numbers vary a lot, take an average of them to get a more accurate idea of the dustfall in your area.) Can you think of reasons why one jar would catch more dust than another? Repeat your investigations in another month, or next year, to see if the amount of dust in the air varies.

4.142 *A thunderstorm experiment*

This experiment requires a watch with a second hand, a pair of compasses, a sheet of paper and, of course, a thunderstorm. As a preliminary, a sketch map of the area within a radius of about 15 kilometres should be drawn to scale on a sheet of paper. Concentric circles should be drawn on the 'map' showing places which are 1, 2, 3 ... 15 kilometres away from the observer. When a thunderstorm occurs, the location of each lightning stroke can be found easily as follows: (i) the direction by visual observation; (ii) the distance by dividing the interval between flash and sound of the thunder in seconds by three—this gives the approximate distance in kilometres. When the storm has passed, you will have a record on your map of the path it took.

Clouds and weather

4.143 *Finding out about clouds*

Clouds are visible evidence of moisture in the air—the more clouds, the more moisture. The moisture may be in the form of liquid water droplets or ice crystals, or both. The type of cloud is an indicator of the stability of the atmosphere in which it forms. Layered clouds (stratiform) indicate generally stable conditions which change rather slowly. Clouds with vertical development (cumuliform) indicate a degree of instability in the atmosphere which produces rapid changes in the clouds. Because of these indications, weather forecasters find it most helpful to have accurate descriptions of the clouds at each weather observing site. Since clouds are continuously in the process of growth or decay they appear in an infinite variety of forms. However, it is possible to define a limited number of characteristic forms generally observed all over the world into which clouds can be broadly grouped (see pages 238 and 239).

In addition to the stratiform and cumuliform types, clouds may be grouped according to the average heights of their bases from the ground as low, middle and high, but this is not a precise classification, since the heights of cloud bases will vary with terrain, available average moisture and weather patterns.

Low clouds include fog, stratus, stratocumulus, cumulus and cumulonimbus. These are found when masses of air move over earth surfaces which are warmer or colder than the air. Uneven heating transferred from the earth's surface to a cooler air layer often causes the formation of cumulus clouds which continue to develop vertically until they become cumulonimbus or thunderheads. The average height of low cloud bases ranges from the earth's surface up to 2,200 metres. Low clouds are usually made up entirely of water droplets and are normally quite dense.

Middle clouds include both altocumulus and altostratus and the average height of their bases will range from 2,200 to 7,700 metres. They are made up of water droplets or ice crystals or both—usually both, and they exhibit considerable varia-

tion in density. An aeroplane pilot flying in a dense water droplet cloud may be able to see only a few metres, while in an ice crystal cloud he may be able to see as far as a kilometre.

High clouds are the cirrus, cirrocumulus and cirrostratus with bases generally above 5,500 metres. They are always made up of ice crystals and vary greatly in density. A distinguishing feature of cirriform clouds is the halo they produce around the sun or moon as a result of refraction of the sunlight or moonlight shining through the ice crystals. Lower clouds (altostratus) containing water droplets exhibit the solar or lunar corona phenomenon rather than the halo.

Another major category of clouds includes 'those exhibiting great vertical development'. This category includes all the low cumulus type, except the fair weather cumulus and stratocumulus. The cumulonimbus or 'thunderhead' is a special category cloud because it may extend through all levels from the very lowest to the very highest and during its life cycle may actually produce nearly all the other cloud types.

The following notes will help you to identify cloud types.

Fog. Fog is a stratiform cloud whose base is at the earth's surface. In mountainous country it is possible to have a single layer of stratus cloud reported as a cloud layer at a valley station and as fog at a mountain observatory. Fog occurs as either water or ice.

Stratus. A layer cloud of the low cloud family which often is the result of fog lifting from the surface. It occurs when the air is stable. Stratus also forms when moist air is lifted up a frontal surface or up sloping terrain, or by advection such as when warm moist air moves over a colder surface. This type of cloud is usually grey in colour and without well-defined outlines. Layers are generally from 100 to 500 metres thick.

Stratocumulus. A layer of clouds whose bases are at a uniform height above the ground, and which displays some tendency to develop vertically. From the ground stratocumulus generally is characterized by light and dark spots or an appearance of furrows or streaks of light and darker areas.

Nimbostratus. This cloud form is always associated with precipitation—drizzle, rain or snow. It is very difficult to judge the height of the base of nimbostratus because it is rather uniformly dull in colour and not sharply defined. It occurs at many levels from very low to middle cloud levels. The precipitation is continuous (not showery) and may range widely in intensity.

Cumulus. Cumulus clouds may be the most common of cloud types and have a very wide range of sizes and shapes. Fair weather cumulus is the ordinary little 'puff ball' clouds of spring and summer skies—they form usually at a uniform height above the ground, grow in size during the hottest part of the day and dissipate towards sundown. A characteristic description is 'a cauliflower appearance'. The edges of cumulus clouds are well defined and sharp. Towering cumulus clouds—sometimes called Mammatocumulus—extend through many thousands of feet and are preliminary to thunderstorms. They grow rapidly in size and give an appearance of boiling, so that their shapes are continuously changing. All cumulus clouds are quite dense and heavy looking. Precipitation from cumuliform clouds occurs as showers rather than as continuous rain or snow. The bases of cumulus clouds may be at nearly any height from very low up to middle levels.

Altocumulus. These are very similar to stratocumulus, but occur at middle levels instead of lower. Some special forms of altocumulus are indicative of particular weather events and have a special significance to weathermen and pilots. One of these is an almond or lens shaped cloud called *lenticularis.* It is associated with wave action in the upper wind field. The individual clouds are continually changing in appearance as cloud forms on one edge and dissipates on the other. Another special form of altocumulus cloud is called *altocumulus castellanus.* It has the appearance of small towering cumulus clouds with towers and turrets continually changing shapes, dissipating and reforming. This type of cloud indicates instability in the middle layers of the atmosphere and often is a sign of possible thunderstorm activity within a few hours.

Altostratus. These are layers of clouds at middle levels of the atmosphere, generally made up of the water droplets or a mixture of water droplets and ice crystals. They are indicative of stable air, and precipitation from this type of cloud is of a light, continuous nature. Altocumulus and altostratus often occur together. The sun or moon shining through altostratus may exhibit the corona effect which distinguishes altostratus from cirrostratus.

Cirrus. These are thin, white, featherlike clouds that often occur in patches or narrow bands. They are composed entirely of ice crystals and their bases are high above the ground. Very often cirrus clouds foretell the approach of a storm frontal system. Cirrus clouds are seldom dense enough to completely hide the sun or moon, although they may thicken enough to make shadows dim and indistinct.

Cirrocumulus. Usually in a layer or sheet, these clouds appear to be tiny white puffs of cotton. They may sometimes be mistaken for altocumulus, but the individual cloud elements are usually much smaller than altocumulus.

Cirrostratus. These are ice crystal clouds at high levels in a sheet or layer that may vary in density from so thin that you have to look very carefully to see them to clouds dense enough to hide the sun. Cirrostratus produces the halo phenomenon. Usually this cloud form signals the approach of a storm frontal system.

Cumulonimbus. This is a heavy-appearing, dense cloud often extending to great heights, which is frequently accompanied by lightning and thunder, heavy showers of rain and sometimes hail and which occasionally produces tornadoes or waterspouts. Characteristic of the cumulonimbus is the spreading out of its top into a long plume or anvil shape. The cumulonimbus is literally a cloud 'factory' as it may during its short life produce nearly all the other types of clouds. Tops of cumulonimbus often extend above 20,000 metres while bases may form anywhere from very near the earth's surface up to 3,000 to 4,000 metres. The growth rate of cumulonimbus is sometimes as much as 2,000 metres per minute.

Upper level

CIRRUS

CIRROCUMULUS

Middle level

ALTOCUMULUS, layer clouds

ALTOCUMULUS, lenticular or almond-shaped cloud

Lower level

NIMBOSTRATUS

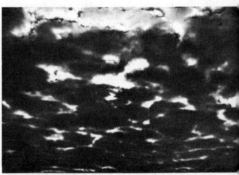

STRATOCUMULUS

Lower to upper levels

CUMULUS, shallow-layer clouds

CUMULUS, towering clouds or 'bubbles'

CIRROSTRATUS

ALTOSTRATUS

STRATUS

CUMULONIMBUS

4.144 *Observations and descriptions of warm and cold fronts*

A. *Warm front.* Warm fronts are preceded by a slowly falling barometer. Cirrus clouds will be observed and precipitation can usually be expected in 24 to 36 hours. The cloud pattern will gradually thicken as it progresses from cirrus to cirrostratus, then altocumulus or altostratus, and finally to nimbostratus or cumulonimbus. Precipitation often begins from dense altostratus clouds before they are obscured by the lower stratus or cumulus types of clouds. As the front passes the wind changes direction, the barometer will rise a little, precipitation will end, the sky will begin to clear, and the temperature will begin to rise noticeably. In summer, afternoon thundershowers may develop behind a warm front.

B. *Cold front.* When a cold front is approaching the barometer will be falling quite rapidly. Cold fronts move faster than warm fronts, having an average speed of 32 to 40 km per hour (although they sometimes move at less than 16 kph and occasionally at more than 56 kph). The procession of cloud types will be proportionately faster than those associated with a warm front. The transition from cirrus to cirrostratus and then to altostratus or altocumulus often takes place within a period of a few hours. Precipitation may start from 12 to 30 hours after the cirrus are first seen.

Level	Approximate altitude		
	Polar climate	Temperate climate	Tropical climate
	km	km	km
Upper	3-8	5-13	6-18
Middle	2-4	2-7	2-8
Lower	Ground level to 2 km	Ground level to 2 km	Ground level to 2 km

Photographs: Météorologie nationale, Paris

In the summer, cumulus clouds will build into cumulonimbus and produce thundershowers. In winter, nimbostratus or stratocumulus will bring rain or snow. When the front passes the wind will shift abruptly, the barometer will rise steadily, and the temperature will fall.

If the front is moving rapidly, clearing will begin quickly, but if the front is comparatively slow moving, cloudiness and some precipitation may last for several hours.

C. *Tornadoes*. Tornadoes are created by the same atmospheric conditions which cause hail and thunderstorms, namely a collision of warm and cold air masses. Tornadoes cannot be predicted, but air conditions which breed them are known, and when these conditions exist weather bureaux usually report 'tornadoes possible'. Covering an area from 70 to 330 metres in width, a tornado usually travels with an average speed of 32 to 63 km per hour, though the wind velocity may be 300 km per hour. In the Northern Hemisphere tornadoes most frequently occur between 1 April and 15 July, and generally in the late afternoon. A tornado is possible whenever the air is humid, with temperature above 26^0 C, and a cold air mass arrives. Mammatocumulus clouds are often seen before and after a tornado.

D. *Hurricanes*. The tropical hurricane is the most devastating of storms. Though occurring all over the world, but under different names, all hurricanes originate in the equatorial regions. North of the equator their general travel direction is N. to NW. to NE. South of the equator hurricanes travel in the opposite direction.

Hurricane cloud formations are much the same as a warm front with a general sequence of changes as follows: (a) cirrus; (b) cirrostratus about 1,600 km in advance of the hurricane; (c) altostratus; (d) nimbostratus rain clouds or cumulonimbus. A halo is often seen about the sun or moon.

Although a hurricane travels only 12 to 24 km per hour, it is accompanied by winds which sometimes reach 240 km an hour. In its life of about ten days a hurricane covers an area of 800 to 3,200 square kilometres. When the barometer begins to rise and the winds change direction the worst of the hurricane is over.

Appendix 1

SI units

	Physical quantity	Name of unit	Symbol	Definition of unit
Basic units:	Amount of substance	mole	mol	
	Electric current	ampere	A	
	Length	metre	m	
	Luminous intensity	candela	cd	
	Mass	kilogramme	kg	
	Thermodynamic temperature	kelvin	K	
	Time	second	s	
Supplementary units:	Plane angle	radian	rad	
	Solid angle	steradian	sr	
Derived units:	Customary temperature	degree Celsius	°C	$t°C = T°K - 273.15$
	Electric capacitance	farad	F	$A^2 s^4 kg^{-1} m^{-2} = A s V^{-1}$
	Electric charge	coulomb	C	$A s$
	Electric potential difference	volt	V	$kg m^2 s^{-3} A^{-1} = J A^{-1} s^{-1}$
	Electric resistance	ohm	Ω	$kg m^2 s^{-3} A^{-2} = V A^{-1}$
	Energy	joule	J	$kg m^2 s^{-2}$
	Force	newton	N	$kg m s^{-2} = J m^{-1}$
	Frequency	hertz	Hz	cycles per second $= s^{-1}$
	Illumination	lux	lx	$cd sr m^{-2}$
	Inductance	henry	H	$kg m^2 s^{-2} A^{-2} = V s A^{-1}$
	Luminous flux	lumen	lm	$cd sr$
	Magnetic flux	weber	Wb	$kg m^2 s^{-2} A^{-1} = V s$
	Magnetic flux density	tesla	T	$kg s^{-2} A^{-1} = V s m^{-2}$
	Power	watt	W	$kg m^2 s^{-3} = J s^{-1}$
	Pressure	pascal	Pa	$kg m^{-1} s^{-2} = N m^{-2}$

	Physical quantity	Name of unit	Symbol	Definition of unit
Units used in conjunction with SI units:	Area	barn	b	10^{-28} m^2
		hectare	ha	10^4 m^2
	Dynamic viscosity	poise	P	10^{-1} kg m^{-1} s^{-1}
	Energy	electronvolt	ev	1.6021×10^{-19} J
	Kinetic viscosity (diffusion)	stokes	St	10^{-4} m^2 s^{-1}
	Length	parsec	pc	3.087×10^{16} m
	Magnetic flux density	gauss	G	10^{-4} T
	Mass	tonne	t	10^3 kg $=$ Mg
	Pressure	bar	bar	10^5 N m^{-2}
	Radioactivity	curie	Ci	3.7×10^{10} s^{-1}
	Volume	litre	l	10^{-3} m^3 $=$ dm^3

Decimal fractions and multiples

Fraction	Prefix	Symbol		Multiple	Prefix	Symbol
10^{-1}	deci	d		10	deca	da
10^{-2}	centi	c		10^2	hecto	h
10^{-3}	milli	m		10^3	kilo	k
10^{-6}	micro	μ		10^6	mega	M
10^{-9}	nano	n		10^9	giga	G
10^{-12}	pico	f		10^{12}	tera	T
10^{-15}	atto	a				

Appendix 2

Conversion chart for units contrary to SI units

Physical quantity	Unit	Equivalent
Area	square inch	6.4516×10^{-4} m^2 = 645.16 mm^2
	square foot	9.2903×10^{-2} m^2
	square yard	8.36127×10^{-1} m^2
	square mile	2.58999 km^2
Density	pound/cubic foot	1.60185×10 kg m^{-3}
Energy	B.Th.U	1.05506×10^3 J
	calorie (15° C)	4.1855 J
	erg	10^{-7} J
	foot-pound-force	1.35582 J
	foot-poundal	4.21401×10^{-1} J
Force	dyne	10^{-5} N
	kilogramme-force	9.80665 N
	pound force	4.44822 N
	poundal	1.38255×10^{-1} N
Length	ångström	10^{-10} m
	inch	2.54×10^{-2} m
	foot	3.048×10^{-1} m
	yard	9.144×10^{-1} m
	mile	1.609344 km
Mass	pound	$4.5359237 \times 10^{-1} \times$ kg
Power	horsepower	7.457×10^2 W
Pressure	atmosphere	1.01325×10^2 kN m^{-2}
	torr	1.33322×10^2 N m^{-2}
	pound-force/square inch	6.89476×10^3 N m^{-2}
Temperature	degree Fahrenheit	$t\,^\circ F = \dfrac{9}{5}\,T\,^\circ C + 32$
Volume	cubic inch	1.63871×10^{-5} m^3
	cubic foot	2.83168×10^{-2} m^3
	U.K. gallon	4.546092×10^{-3} m^3
	U.S. gallon	3.788404×10^{-3} m^3

Appendix 3

Periodic table

1	1 1.008 H Hydrogen									
2	3 6.940 Li Lithium	4 9.013 Be Beryllium								
3	11 22.997 Na Sodium	12 24.32 Mg Magnesium								
4	19 39.096 K Potassium	20 40.08 Ca Calcium	21 44.96 Sc Scandium	22 47.90 Ti Titanium	23 50.95 V Vanadium	24 52.01 Cr Chromium	25 54.93 Mn Manganese	26 55.85 Fe Iron	27 58.94 Co Cobalt
5	37 85.48 Rb Rubidium	38 87.63 Sr Strontium	39 88.92 Y Yttrium	40 91.22 Zr Zirconium	41 92.91 Nb Niobium	42 95.95 Mo Molybdenum	43 (99) Tc Technetium	44 101.7 Ru Ruthenium	45 102.91 Rh Rhodium
6	55 132.91 Cs Cesium	56 137.36 Ba Barium	57-71 Rare earth elements	72 178.6 Hf Hafnum	73 180.88 Ta Tantalum	74 183.92 W Tungsten	75 186.31 Re Rhenium	76 190.2 Os Osmium	77 193.1 Ir Iridium
7	87 (223) Fr Francium	88 226.05 Ra Radium	89-98 Actinide series						

57 138.92 La Lanthanum	58 140.13 Ce Cerium	59 140.92 Pr Praseo- dymium	60 144.27 Nd Neodymium	61 (145) Pm Promethium	62 150.43 Sm Samarium
89 227 Ac Actinium	90 232.12 Th Thorium	91 (231) Pa Protactinium	92 238.07 U Uranium	93 (237) Np Neptunium	94 (242) Pu Plutonium

Key

Atomic number — 12 24.32 — Atomic weight

Mg ← Symbol

Magnesium

Name

								2 4.003 He Helium
			5 10.82 B Boron	6 12.01 C Carbon	7 14.008 N Nitrogen	8 16.00 O Oxygen	9 19.00 F Fluorine	10 20.183 Ne Neon
			13 26.98 Al Aluminium	14 28.09 Si Silicon	15 30.98 P Phosphorus	16 32.066 S Sulphur	17 35.457 Cl Chlorine	18 39.9 A Argon
28 58.69 Ni Nickel	29 63.54 Cu Copper	30 65.38 Zn Zinc	31 69.72 Ga Gallium	32 72.60 Ge Germanium	33 74.91 As Arsenic	34 78.96 Se Selenium	35 79.916 Br Bromine	36 83.6 Kr Krypton
46 106.7 Pd Palladium	47 107.88 Ag Silver	48 112.41 Cd Cadmium	49 114.76 In Indium	50 118.70 Sn Tin	51 121.75 Sb Antimony	52 127.60 Te Tellurium	53 126.904 I Iodine	54 131.30 Xe Xenon
78 195.23 Pt Platinum	79 197.2 Au Gold	80 200.61 Hg Mercury	81 204.39 Tl Thallium	82 207.21 Pb Lead	83 209 Bi Bismuth	84 210 Po Polonium	85 (210) At Astitine	86 222 Rn Radon
63 152.0 Eu Europium	64 156.9 Gd Gadolinium	65 159.2 Tb Terbium	66 162.46 Du Dysprosium	67 164.94 Ho Holmium	68 167.2 Er Erbium	69 169.4 Tm Thulium	70 173.04 Yb Ytterbium	71 174.99 Lu Lutetium
95 (243) Am Americium	96 (243) Cm Curium	97 (245) Bk Berkelium	98 (246) Cf Californium	99 (254) Es Einsteinium	100 (253) Fm Fermium	101 (256) Md Men- delevium	102 (254) No Nobelium	103 (257) Lw Lawrencium

Appendix 4

A table of the elements

Symbol	Atomic No.	Name	Interesting facts
Ac	89	Actinium	Radioactive metal; rare
Al	13	Aluminium	Alloys are strong and lightweight
Am	95	Americium	Made by man; highly radioactive
Sb	51	Antimony	Silvery, brittle metal; important alloy is type metal
Ar	18	Argon	Colourless gas in air; used in electric light bulbs
As	33	Arsenic	Grey solid; compounds are poisonous
At	85	Astatine	Made by man from bismuth; radioactive non-metal
Ba	56	Barium	Lightweight metal; soft, silvery white
Bk	97	Berkelium	Made by man (1950); highly radioactive metal
Be	4	Beryllium	Lightweight metal; small beryllium-copper springs are very long lived
Bi	83	Bismuth	Silvery pink metal; makes hard alloys with low melting point
B	5	Boron	Solid non-metal; present in borax
Br	35	Bromine	Red liquid; name means 'stench'
Cd	48	Cadmium	Silvery metal; often electroplated over other metals
Ca	20	Calcium	Lightweight metal; compounds are abundant in earth's crust
Cf	98	Californium	Made by man (1950); highly radioactive metal
C	6	Carbon	Key element of organic chemistry—in all plants and animals
Ce	58	Cerium	Hard metal; makes sparks for lighters
Cs	55	Cesium	Soft, silvery metal; melts in boiling water
Cl	17	Chlorine	Greenish-yellow gas; poisonous; good bleaching agent
Cr	24	Chromium	Bright, silvery metal; used in stainless steel alloys
Co	27	Cobalt	Silvery metal; part of powerful magnetic alloy
Cu	29	Copper	Red metal; conductor of electricity
Cm	96	Curium	Made by man from plutonium; highly radioactive metal
Dy	66	Dysprosium	Rare earth metal; name means 'hard to get at'
Es	99	Einsteinium	Made by bombarding uranium with the nuclei of nitrogen atoms; atomic weight 247; highly radioactive
Er	68	Erbium	Rare earth metal
Eu	63	Europium	Rare earth metal
Fm	100	Fermium	Made by adding neutrons to plutonium to make californium and then more neutrons to make element 100; highly radioactive
F	9	Fluorine	Highly active and poisonous gas
Fr	87	Francium	Radioactive metal; extremely rare; also produced by nuclear reactions
Gd	64	Gadolinium	Rare earth metal; free element not yet prepared
Ga	31	Gallium	Shining, white metal; usually separated from zinc ores
Ge	32	Germanium	Grey brittle metal; similar to tin
Au	79	Gold	Long famous for decoration and money standard
Hf	72	Hafnium	Heavy metal; similar to zirconium
Hc	2	Helium	Chemically inactive gas twice as heavy as hydrogen
Ho	67	Holmium	Rare earth metal; free element not yet produced

Symbol	Atomic No.	Name	Interesting facts
H	1	Hydrogen	Colourless gas, lightest of all
In	49	Indium	Soft, silvery metal; similar to aluminium
I	53	Iodine	Brownish-black solid; when heated changes to beautiful purple vapour
Ir	77	Iridium	Silvery metal; alloyed with platinum for pen points
Fe	26	Iron	Second most abundant metal
Kr	36	Krypton	Inert colourless gas in air
La	57	Lanthanum	Rare earth metal
Lw	103	Lawrencium	Newest element, extremely short life, radioactive, man made (1961)
Pb	82	Lead	Heavy, bluish-white, soft metal
Li	3	Lithium	Lightest metal known; soft
Lu	71	Lutetium	Rare earth metal; of little use
Mg	12	Magnesium	Combines lightness with strength
Mn	25	Manganese	Heavy metal; highly important in steel industry
Md	101	Mendelevium	Short lived, highly radioactive
Hg	80	Mercury	Heavy, silvery, liquid metal
Mo	42	Molybdenum	Silvery metal; has many important steel alloys
Nd	60	Neodymium	Rare earth metal; compounds pink
Ne	10	Neon	Inert gas in the air; makes brilliant electric lights
Np	93	Neptunium	Made by man from uranium; radioactive
Ni	28	Nickel	Makes tough corrosion-resistant steel
Nb	41	Niobium	Silvery metal; formerly called columbium
N	7	Nitrogen	Colourless gas; makes up 78% of air
No	102	Nobelium	Short lived; highly radioactive
Os	76	Osmium	Silvery metal; heaviest element
O	8	Oxygen	Colourless gas; abundant element
Pd	46	Palladium	Resembles platinum
P	15	Phosphorous	Soft, non-metallic solid; ignites easily
Pt	78	Platinum	Silvery metal; useful for laboratory vessels and instruments
Pu	94	Plutonium	Made by man; highly important for nuclear fission
Po	84	Polonium	Radioactive metal; found by the Curies just before radium
K	19	Potassium	Soft metal; lighter than water
Pr	59	Praseodymium	Rare earth metal
Pm	61	Promethium	Rare earth metal; also made by man from praseodymium
Pa	91	Protactinium	Radioactive metal; present in all uranium ores
Ra	88	Radium	Radioactive metal; discovery stimulated research on radioactivity
Rn	86	Radon	Heaviest gas known; radioactive; comes from radium
Re	75	Rhenium	Heavy metal; resembles manganese
Rh	45	Rhodium	Heavy metal; looks like aluminium; used for electroplating jewellery
Rb	37	Rubidium	Soft metal; rare; highly active chemically
Ru	44	Ruthenium	Hard, grey, brittle metal

Symbol	Atomic No.	Name	Interesting facts
Sm	62	Samarium	Rare earth metal
Sc	21	Scandium	Free element not yet produced; rare
Se	34	Selenium	Solid non-metal; resembles sulphur in chemical changes
Si	14	Silicon	Solid non-metal; second in abundance
Ag	47	Silver	Best conductor of heat and electricity
Na	11	Sodium	Soft, highly active metal; lighter than water
Sr	38	Strontium	Hard, active metal; resembles calcium chemically
S	16	Sulphur	Solid yellow non-metal
Ta	73	Tantalum	Looks like polished iron; makes alloy steel
Tc	43	Technetium	Heavy metal; found among fission products or uranium
Te	52	Tellurium	Solid non-metal; resembles sulphur in chemical changes
Tb	65	Terbium	Rare earth metal; free metal not yet prepared
Tl	81	Thallium	Solid metal; resembles lead; its salts are very poisonous
Th	90	Thorium	Heavy, grey metal; all its compounds are radioactive
Tm	69	Thulium	Rare earth metal not yet obtained as free element
Sn	50	Tin	Silvery metal; electroplated on steel for cans
Ti	22	Titanium	Strong, hard metal; new production methods make future bright
U	92	Uranium	Object of world-wide search because of need for nuclear fission
V	23	Vanadium	Grey metal; difficult to melt; makes strong, tough alloy steel
W	74	Tungsten	Heavy metal; formerly called wolfram; has highest melting point
Xe	54	Xenon	Rare, inert, colourless gas in the air
Yb	70	Ytterbium	Rare earth metal
Y	39	Yttrium	Rare earth metal; more abundant than many other rare earth metals
Zn	30	Zinc	Bluish-white metal; used for outer coat on galvanized iron
Zr	40	Zirconium	Gold-coloured metal; the compound, zircon is a gem

Appendix 5

Acid-base indicators

Indicator	pH range	Quantity of indicator per 10 ml	Colour	
			Acid	Alkaline
Thymol blue	1.2–2.8	1–2 drops 0.1% solution in water	Red	Yellow
Tropeolin OO	1.3–3.2	1 drop 1% solution in water	Red	Yellow
Methyl yellow (B)	2.9–4.0	1 drop 0.1% solution in 90% alcohol	Red	Yellow
Methyl orange (B)	3.1–4.4	1 drop 0.1% solution in water	Red	Orange
Brom phenol blue (A)	3.0–4.6	1 drop 0.1% solution in water	Yellow	Blue-violet
Brom cresol green	4.0–5.6	1 drop 0.1% solution in water	Yellow	Blue
Methyl red (A)	4.4–6.2	1 drop 0.1% solution in water	Red	Yellow
Brom thymol blue	6.2–7.6	1 drop 0.1% solution in water	Yellow	Blue
Phenol red (A)	6.4–8.0	1 drop 0.1% solution in water	Yellow	Red
Neutral red (B)	6.8–8.0	1 drop 0.1% solution in 70% alcohol	Red	Yellow
Thymol blue	8.0–9.6	1–5 drops 0.1% solution in water	Yellow	Blue

Indicator	pH range	Quantity of indicator per 10 ml	Colour	
			Acid	Alkaline
Phenolphthalein (A)	9.0–11.0	1–5 drops 0.1% solution in 90% alcohol	Colourless	Red
Thymolphthalein	9.4–10.6	1 drop 0.1% solution in 90% alcohol	Colourless	Blue
Alizarin yellow	10.0–12.0	1–5 drops 0.1% solution in 90% alcohol	Yellow	Orange-brown
Tropeolin O	11.0–13.0	1 drop 0.1% solution in water	Yellow	Orange-brown
Nitramine (B)	11.0–13.0	1–2 drops 0.1% solution in 70% alcohol	Colourless	Orange-brown
Trinitrobenzoic acid	12.0–13.4	1 drop 0.1% solution in water	Colourless	Orange-red

Appendix 6

Relative humidity (percentage)—°C

Temperature of dry bulb (°C)	Depression of the wet bulb (°C)														
	1	2	3	4	5	6	7	8	9	10	12	14	16	18	20
50	94	89	84	79	74	70	65	61	57	53	46	40	33	28	22
45	94	88	83	78	73	68	63	59	55	51	42	35	28	22	16
40	93	88	82	77	71	65	61	56	52	47	38	31	23	16	10
35	93	87	80	75	68	62	57	52	47	42	33	24	16	8	
30	92	86	78	72	65	59	53	47	41	36	26	16	8		
25	91	84	76	69	61	54	47	41	35	29	17	6			
20	90	81	73	64	56	47	40	32	26	18	5				
15	89	79	68	59	49	39	30	21	12	4					
10	87	75	62	51	38	27	17	5							

Appendix 7

Equivalent temperatures in different scales

	Kelvin	Celsius	Fahrenheit
Absolute zero	0° K	−273° C	−459° F
Fahrenheit zero	255° K	−18° C	0° F
Freezing point of water	273° K	0° C	32° F
Boiling point of water	373° K	100° C	212° F

Appendix 8

Logarithms

	0	1	2	3	4	5	6	7	8	9	1	2	3	4	5	6	7	8	9
10	0000	0043	0086	0128	0170						4	9	13	17	21	26	30	34	38
						0212	0253	0294	0334	0374	4	8	12	16	20	24	28	32	37
11	0414	0453	0492	0531	0569						4	8	12	15	19	23	27	31	35
						0607	0645	0682	0719	0755	4	7	11	15	19	22	26	30	33
12	0792	0828	0864	0899	0934	0969					3	7	11	14	18	21	25	28	32
							1004	1038	1072	1106	3	7	10	14	17	20	24	27	31
13	1139	1173	1206	1239	1271						3	7	10	13	16	20	23	26	30
						1303	1335	1367	1399	1430	3	7	10	12	16	19	22	25	29
14	1461	1492	1523	1553							3	6	9	12	15	18	21	24	28
					1584	1614	1644	1673	1703	1732	3	6	9	12	15	17	20	23	26
15	1761	1790	1818	1847	1875	1903					3	6	9	11	14	17	20	23	26
							1931	1959	1987	2014	3	5	8	11	14	16	19	22	25
16	2041	2068	2095	2122	2148						3	5	8	11	14	16	19	22	24
						2175	2201	2227	2253	2279	3	5	8	10	13	15	18	21	23
17	2304	2330	2355	2380	2405	2430					3	5	8	10	13	15	18	20	23
							2455	2480	2504	2529	2	5	7	10	12	15	17	19	22
18	2553	2577	2601	2625	2648						2	5	7	9	12	14	16	19	21
						2672	2695	2718	2742	2765	2	5	7	9	11	14	16	18	21
19	2788	2810	2833	2856	2878						2	4	7	9	11	13	16	18	20
						2900	2923	2945	2967	2989	2	4	6	8	11	13	15	17	19
20	3010	3032	3054	3075	3096	3118	3139	3160	3181	3201	2	4	6	8	11	13	15	17	19
21	3222	3243	3263	3284	3304	3324	3345	3365	3385	3404	2	4	6	8	10	12	14	16	18
22	3424	3444	3464	3483	3502	3522	3541	3560	3579	3598	2	4	6	8	10	12	14	15	17
23	3617	3636	3655	3674	3692	3711	3729	3747	3766	3784	2	4	6	7	9	11	13	15	17
24	3802	3820	3838	3856	3874	3892	3909	3927	3945	3962	2	4	5	7	9	11	12	14	16
25	3979	3997	4014	4031	4048	4065	4082	4099	4116	4133	2	3	5	7	9	10	12	14	15
26	4150	4166	4183	4200	4216	4232	4249	4265	4281	4298	2	3	5	7	8	10	11	13	15
27	4314	4330	4346	4362	4378	4393	4409	4425	4440	4456	2	3	5	6	8	9	11	13	14
28	4472	4487	4502	4518	4533	4548	4564	4579	4594	4609	2	3	5	6	8	9	11	12	14
29	4624	4639	4654	4669	4683	4698	4713	4728	4742	4757	1	3	4	6	7	9	10	12	13
30	4771	4786	4800	4814	4829	4843	4857	4871	4886	4900	1	3	4	6	7	9	10	11	13
31	4914	4928	4942	4955	4969	4983	4997	5011	5024	5038	1	3	4	6	7	8	10	11	12
32	5051	5065	5079	5092	5105	5119	5132	5145	5159	5172	1	3	4	5	7	8	9	11	12
33	5185	5198	5211	5224	5237	5250	5263	5276	5289	5302	1	3	4	5	6	8	9	10	12
34	5315	5328	5340	5353	5366	5378	5391	5403	5416	5428	1	3	4	5	6	8	9	10	11
35	5441	5453	5465	5478	5490	5502	5514	5527	5539	5551	1	2	4	5	6	7	9	10	11
36	5563	5575	5587	5599	5611	5623	5635	5647	5658	5670	1	2	4	5	6	7	8	10	11
37	5682	5694	5705	5717	5729	5740	5752	5763	5775	5786	1	2	3	5	6	7	8	9	10
38	5798	5809	5821	5832	5843	5855	5866	5877	5888	5899	1	2	3	5	6	7	8	9	10
39	5911	5922	5933	5944	5955	5966	5977	5988	5999	6010	1	2	3	4	5	7	8	9	10
40	6021	6031	6042	6053	6064	6075	6085	6096	6107	6117	1	2	3	4	5	6	8	9	10
41	6128	6138	6149	6160	6170	6180	6191	6201	6212	6222	1	2	3	4	5	6	7	8	9
42	6232	6243	6253	6263	6274	6284	6294	6304	6314	6325	1	2	3	4	5	6	7	8	9
43	6335	6345	6355	6365	6375	6385	6395	6405	6415	6425	1	2	3	4	5	6	7	8	9
44	6435	6444	6454	6464	6474	6484	6493	6503	6513	6522	1	2	3	4	5	6	7	8	9
45	6532	6542	6551	6561	6571	6580	6590	6599	6609	6618	1	2	3	4	5	6	7	8	9
46	6628	6637	6646	6656	6665	6675	6684	6693	6702	6712	1	2	3	4	5	6	7	7	8
47	6721	6730	6739	6749	6758	6767	6776	6785	6794	6803	1	2	3	4	5	5	6	7	8
48	6812	6821	6830	6839	6848	6857	6866	6875	6884	6893	1	2	3	4	4	5	6	7	8
49	6902	6911	6920	6928	6937	6946	6955	6964	6972	6981	1	2	3	4	4	5	6	7	8
50	6990	6998	7007	7016	7024	7033	7042	7050	7059	7067	1	2	3	3	4	5	6	7	8

Note: These tables are so constructed that the fourth figure of a logarithm obtained by their use is never more than one unit above or below the best 4-figure approximation. E.g. if the logarithm found is 0.5014 the best 4-figure approximation may be 0.5013, 0.5014 or 0.5015. Greater accuracy than this cannot be obtained by the use of a uniform table of differences of this kind.

	0	1	2	3	4	5	6	7	8	9	1	2	3	4	5	6	7	8	9
51	7076	7084	7093	7101	7110	7118	7126	7135	7143	7152	1	2	3	3	4	5	6	7	8
52	7160	7168	7177	7185	7193	7202	7210	7218	7226	7235	1	2	2	3	4	5	6	7	7
53	7243	7251	7259	7267	7275	7284	7292	7300	7308	7316	1	2	2	3	4	5	6	6	7
54	7324	7332	7340	7348	7356	7364	7372	7380	7388	7396	1	2	2	3	4	5	6	6	7
55	7404	7412	7419	7427	7435	7443	7451	7459	7466	7474	1	2	2	3	4	5	5	6	7
56	7482	7490	7497	7505	7513	7520	7528	7536	7543	7551	1	2	2	3	4	5	5	6	7
57	7559	7566	7574	7582	7589	7597	7604	7612	7619	7627	1	2	2	3	4	5	5	6	7
58	7634	7642	7649	7657	7664	7672	7679	7686	7694	7701	1	1	2	3	4	4	5	6	7
59	7709	7716	7723	7731	7738	7745	7752	7760	7767	7774	1	1	2	3	4	4	5	6	7
60	7782	7789	7796	7803	7810	7818	7825	7832	7839	7846	1	1	2	3	4	4	5	6	6
61	7853	7860	7868	7875	7882	7889	7896	7903	7910	7917	1	1	2	3	4	4	5	6	6
62	7924	7931	7938	7945	7952	7959	7966	7973	7980	7987	1	1	2	3	3	4	5	6	6
63	7993	8000	8007	8014	8021	8028	8035	8041	8048	8055	1	1	2	3	3	4	5	5	6
64	8062	8069	8075	8082	8089	8096	8102	8109	8116	8122	1	1	2	3	3	4	5	5	6
65	8129	8136	8142	8149	8156	8162	8169	8176	8182	8189	1	1	2	3	3	4	5	5	6
66	8195	8202	8209	8215	8222	8228	8235	8241	8248	8254	1	1	2	3	3	4	5	5	6
67	8261	8267	8274	8280	8287	8293	8299	8306	8312	8319	1	1	2	3	3	4	5	5	6
68	8325	8331	8338	8344	8351	8357	8363	8370	8376	8382	1	1	2	3	3	4	4	5	6
69	8388	8395	8401	8407	8414	8420	8426	8432	8439	8445	1	1	2	2	3	4	4	5	6
70	8451	8457	8463	8470	8476	8482	8488	8494	8500	8506	1	1	2	2	3	4	4	5	6
71	8513	8519	8525	8531	8537	8543	8549	8555	8561	8567	1	1	2	2	3	4	4	5	5
72	8573	8579	8585	8591	8597	8603	8609	8615	8621	8627	1	1	2	2	3	4	4	5	5
73	8633	8639	8645	8651	8657	8663	8669	8675	8681	8686	1	1	2	2	3	4	4	5	5
74	8692	8698	8704	8710	8716	8722	8727	8733	8739	8745	1	1	2	2	3	4	4	5	5
75	8751	8756	8762	8768	8774	8779	8785	8791	8797	8802	1	1	2	2	3	3	4	5	5
76	8808	8814	8820	8825	8831	8837	8842	8848	8854	8859	1	1	2	2	3	3	4	5	5
77	8865	8871	8876	8882	8887	8893	8899	8904	8910	8915	1	1	2	2	3	3	4	4	5
78	8921	8927	8932	8938	8943	8949	8954	8960	8965	8971	1	1	2	2	3	3	4	4	5
79	8976	8982	8987	8993	8998	9004	9009	9015	9020	9025	1	1	2	2	3	3	4	4	5
80	9031	9036	9042	9047	9053	9058	9063	9069	9074	9079	1	1	2	2	3	3	4	4	5
81	9085	9090	9096	9101	9106	9112	9117	9122	9128	9133	1	1	2	2	3	3	4	4	5
82	9138	9143	9149	9154	9159	9165	9170	9175	9180	9186	1	1	2	2	3	3	4	4	5
83	9191	9196	9201	9206	9212	9217	9222	9227	9232	9238	1	1	2	2	3	3	4	4	5
84	9243	9248	9253	9258	9263	9269	9274	9279	9284	9289	1	1	2	2	3	3	4	4	5
85	9294	9299	9304	9309	9315	9320	9325	9330	9335	9340	1	1	2	2	3	3	4	4	5
86	9345	9350	9355	9360	9365	9370	9375	9380	9385	9390	1	1	2	2	3	3	4	4	5
87	9395	9400	9405	9410	9415	9420	9425	9430	9435	9440	0	1	1	2	2	3	3	4	4
88	9445	9450	9455	9460	9465	9469	9474	9479	9484	9489	0	1	1	2	2	3	3	4	4
89	9494	9499	9504	9509	9513	9518	9523	9528	9533	9538	0	1	1	2	2	3	3	4	4
90	9542	9547	9552	9557	9562	9566	9571	9576	9581	9586	0	1	1	2	2	3	3	4	4
91	9590	9595	9600	9605	9609	9614	9619	9624	9628	9633	0	1	1	2	2	3	3	4	4
92	9638	9643	9647	9652	9657	9661	9666	9671	9675	9680	0	1	L	2	2	3	3	4	4
93	9685	9689	9694	9699	9703	9708	9713	9717	9722	9727	0	1	1	2	2	3	3	4	4
94	9731	9736	9741	9745	9750	9754	9759	9763	9768	9773	0	1	1	2	2	3	3	4	4
95	9777	9782	9786	9791	9795	9800	9805	9809	9814	9818	0	1	1	2	2	3	3	4	4
96	9823	9827	9832	9836	9841	9845	9850	9854	9859	9868	0	1	1	2	2	3	3	4	4
97	9868	9872	9877	9881	9886	9890	9894	9899	9903	9908	0	1	1	2	2	3	3	4	4
98	9912	9917	9921	9926	9930	9934	9939	9943	9948	9952	0	1	1	2	2	3	3	4	4
99	9956	9961	9965	9969	9974	9978	9983	9987	9991	9996	0	1	1	2	2	3	3	3	4

Appendix 9

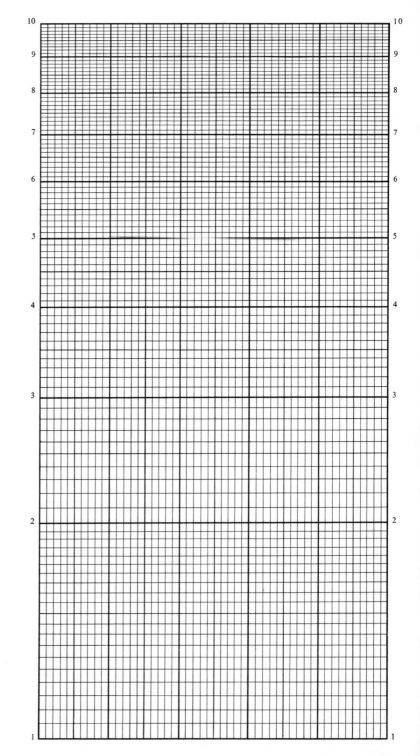

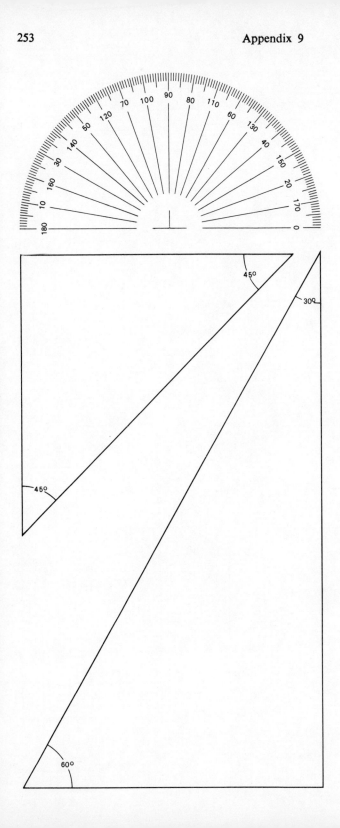

Index